Energy Technology and Directions for the Future

Editor

Vikas Jasiwal

scitus
academics

Energy Technology and Directions for the Future
Edited by **Vikas Jasiwal**

Printed in 2017

ISBN: 978-1-68117-361-0

Library of Congress Control Number: 2015941552

© 2016 by
SCITUS Academics LLC,
616, Corporate Way, Suite 2, 4766,
Valley Cottage, NY 10989

www.scitusacademics.com

Contents

Preface

Energy: Technology and Directions for the Future presents the fundamentals of energy for scientists and engineers. It is a survey of energy sources that will be available for use in the 21st century energy mix. The reader will learn about the history and science of several energy sources as well as the technology and social significance of energy. Themes in the book include thermodynamics, electricity distribution, geothermal energy, fossil fuels, solar energy, nuclear energy, alternate energy (wind, water, biomass), energy and society, energy and the environment, sustainable development, the hydrogen economy, and energy forecasting. Energy: Technology and Directions for the Future presents the fundamen-tals of energy for scientists and engineers. It recognizes that society's dependence on fossil energy in the early twenty-first century is in a state of transition to a broader energy mix. Forecasts of the twenty-first-century energy mix show that a range of scenarios is possible.

Editor

Chapter 1

Carbon Dioxide Separation from Flue Gases: A Technological Review Emphasizing Reduction in Greenhouse Gas Emissions

Mohammad Songolzadeh[1], Mansooreh Soleimani[1], Maryam Takht Ravanchi[2], and Reza Songolzadeh[3]

[1]Department of Chemical Engineering, Amirkabir University of Technology, Tehran, Iran

[2]Catalyst Research Group, Petrochemical Research and Technology Company, National Petrochemical Company, Tehran, Iran

[3]Department of Petroleum Engineering, Petroleum University of Technology, Ahwaz, Iran

ABSTRACT

Increasing concentrations of greenhouse gases (GHGs) such as CO_2 in the atmosphere is a global warming. Human activities are a major cause of increased CO_2 concentration in atmosphere, as in recent decade, two-third of greenhouse effect was caused by human activities. Carbon capture and storage (CCS) is a major strategy that can be used to reduce GHGs emission. There are three methods for CCS: pre-combustion capture, oxy-fuel process, and post-combustion capture. Among them, post-combustion capture is the most important one because it offers flexibility and it can be easily added to the operational units. Various technologies are used for CO_2 capture, some of them include: absorption, adsorption, cryogenic distillation, and membrane separation. In this paper, various technologies for post-combustion are compared and the best condition for using each technology is identified.

INTRODUCTION

There are ten primary GHGs including water vapor (H_2O), carbon dioxide (CO_2), methane (CH_4), and nitrous oxide (N_2O) that are naturally occurring. Perfluorocarbons (CF_4, C_2F_6), hydrofluorocarbons (CHF_3, CF_3CH_2F, and CH_3CHF_2), and sulfur hexafluoride (SF_6), are only present in the atmosphere due to industrial processes. Water vapor is the most important, abundant and dominant greenhouse gas, and CO_2 is the second-most important one (Table 1). Concentration of water vapor depends on temperature and other meteorological conditions, and not directly upon human activities. So it was not indicated in Table 1 [1–3].

Table 1: The main greenhouse gases and their concentration [2, 3]

Compound	Preindustrial concentration (ppmv)	Concentration in 2011 (ppmv)	Atmospheric lifetime (years)	Main human activity source	GWP**

Carbon dioxide (CO_2)	280	388.5	~100	Fossil fuels, cement production, land use	1
Methane (CH_4)	0.715	1.87/1.748	12	Fossil fuels, rice paddies, waste dumps, livestock	25
Nitrous oxide (N_2O)	0.27	0.323	114	Fertilizers, combustion industrial processes	298
CFC-12 (CCL_2F_2)	0	0.000533	100	Liquid coolants, foams	10,900
CF-113 (CCl_2CClF_2)	0	0.00000075	85	n.a.	6,130
HFC 23 (CHF_3)	0	0.000018	270	Electronics, refrigerants	11,700
HCFC-22 (CCl_2F_2)	0	0.000218	12	Refrigerants	1,810
HFC 134 (CF_3CH_2F)	0	0.000035	14	Refrigerants	1,300
HCFC-141b (CH_3CCl_2F)	0	0.00000022	9.3	n.a.	725
HCFC-142b (CH_3CClF_2)	0	0.00000020	17.9	n.a.	2,310
HFC 152 (CH_3CHF_2)	0	0.0000039	1.4	Industrial processes	140
Perfluoromethane (CF_4)	0.00004	0.00008*	50,000	Aluminum production	6,500
Perfluoroethane (C_2F_6)	0	0.000003*	10,000	Aluminum production	9,200
Sulfur hexafluoride (SF_6)	0	0.00000712*	3,200	Dielectric fluid	22,800

*Concentration in 2011.

**Global warming potentials (GWPs) measure the relative effectiveness of GHGs in trapping the Earth's heat.

high chemical and thermal stability and should be harmless for labor persons [51–53].

The solvents used for CO_2 absorption can be divided into two categories: physical and chemical solvents. Physical solvent processes use organic solvents to physically absorb acid gas components rather than reacting chemically, but chemical absorption depends on acid-base neutralization reactions using alkaline solvents [54,55]. In the recent years, many studies have compared the performance of different solvents as listed in Table 2.

Table 2: Various solvents suggested for CO_2 separation

Group of solvents	Advantage	Disadvantage	Application	Reference
Physical				
Dimethyl ether of polyethylene glycol (Selexol)	(i) Require low energy for re-generation (less than 20% of the value for chemical absorbent) (ii) Low vapor pressure, low toxicity, and less corrosive solvent	(i) Dependent on temperature and pressure; therefore they are not suitable for post-combustion process (ii) Low capacity for CO_2 absorption	Natural gas sweetening	[29, 39,57, 62, 63]
Glycol			Capturing CO_2 and H_2S at higher concentration	
Glycol carbonate			Separating CO_2 from other gases	
Methanol (Rectisol)			CO_2 removal from various streams	
Fluorinated solvent			(i) CO_2 removal from various streams (ii) Separating CO_2 from other gases	
Chemical				

Alkanolamines: monoethanolamine (MEA), diethanolamine (DEA), and methyl diethanolamine (MDEA)	(i) React rapidly (ii) High selectively (between acid and other gases) (iii) Reversible absorption process (iv) Inexpensive solvent	(i) Low CO_2 loading capacity (ii) Solvent degradation in existence of SO_2 and O_2 in flue gas (concentrations must be less than 10 ppm and 1 ppm) (iii) High equipment corrosion rate (iv) High energy consumption	Important for removing acidic components from gas streams	[58, 60,61,64–66]
Amino acid and aqueous amino acid salt	(i) The possibility of adding a salt functional group. (ii) The nonvolatility of solvents (iii) Having high surface tension (iv) Having better resistance to degradation than other chemical solvents (v) Better performance than MEA of the same concentration for CO_2 absorption	Decreased performance in the presence of oxygen	Suggested for CO_2 separation from flue gases	[65,67–69]
Ammonia	(i) No degradation in the presence of SO_2 and O_2 in the flue gases (ii) No corrosion effect (iii) Require low energy to regeneration (1/3 that required with MEA) (iv) Low costs with aqueous ammonia, respectively, 15% and 20% less than with MEA	(i) Reversible at lower temperatures (not suitable for post-combustion) (ii) Production of solid products and their operating problems (iii) Explosion of dry CO_2-NH_3 reaction in the high concentration of CO_2 in the flue gas (explosive limit for NH_3 gas is 15–28%)	Suggested for CO_2 separation from flue gases	[39, 70]

Ionic liquid (IL)	(i) Very low vapor pressure (ii) Good thermal stability (iii) High polarity (iv) Nontoxicity	Increased viscosity with CO_2 absorption	Suggested for CO_2 separation from flue gases	[71–74]
Aqueous piperazine (PZ)	(i) Fast absorption kinetics (CO_2 absorption rate with aqueous PZ is more than double that of MEA) (ii) Low degradation rates for CO_2 separation (iii) Negligible thermal degradation in concentrated PZ solutions (iv) Favorable equilibrium characteristics (v) Very low heat of absorption (10–15 kCal/mol CO_2), 80–90% energy required for aqueous amine system	Lower oxidative degradation of concentrated PZ (i.e., 4 times slower than MEA in the presence of the combination of $Fe^{2+}/Cr^{3+}/Ni^{2+}$ and Fe^{2+}/V^{5+})	(i) Effective for treating syngas at high temperatures (ii) Application of additional amine promoters for natural gas treating and CO_2 separation from flue gases	[29, 66,75, 76]

(1) *Alkanolamines.* Between various solvent groups, alkanolamines group is the most important and more used for CO_2 separation. A major problem in the usage of amines for CO_2 absorption is equipment corrosion, so Albritton et al. [56] examined corrosion rate of various amine solvents and suggested corrosion rate could reduce in the following order: monoethanolamine (MEA) > 2-amino-2-methyl-1-propanol (AMP) > diethanolamine (DEA) > methyl diethanolamine (MDEA).

On the other way, MEA can react more quickly with CO_2 than MDEA, but MDEA has higher CO_2 absorption capacity and requires lower energy to regenerate CO_2 [39, 57, 58]. Thus, it can be concluded that MEA is one of the best amine solvents for CO_2 separation. Idem et al. [59] reported substantial reduction in energy requirements and modest reduction in circulation rates for amine blends relative to the corresponding single amine system of similar total amine concentration. Wang et al. [57] found that when MEA and MDEA are mixed at the appropriate ratio, the energy consumption for CO_2 regeneration is reduced significantly. Dave et al. [28] compared the performance of several amine solvents and ammonia solutions at various concentrations. They showed that 30 wt% AMP based process has the lowest overall energy requirement among the solvents considered in their study (30% MEA, 30% MDEA, 2.5% NH_3, and 5% NH_3) [28, 60].

Knudsen et al. [61] studies showed that it is possible to run the post-combustion capture plant continuously while achieving roughly 90% CO_2 separation levels and CASTOR-2 (blended amine solvents), operated in pilot scale with lower steam requirement and liquid-to-gas ratio (L/G) than the conventional MEA solvent.

Besides alkanolamines, carbonate-bicarbonate buffers and hindered amines are used in the bulk removal of CO_2 owing to the low steam requirement for its regeneration. Mitsubishi Heavy Industries and Kansai Electric have employed other patented chemical solvents—strictly hindered amines called KS-1, KS-2, or KS-3. The regeneration heat of KS solvents is said to be ~3 GJ/t CO_2, that is, 20% lower than that of MEA with ~3.7 GJ/t CO_2 [60, 64, 77]. Generally, the overall cost of amine absorption/stripping technology for CO_2 capture process is 52–77 US$/ton CO_2 [71].

(2) *Amino Acid.* Amino acids have the same functional groups as alkanolamines and can be expected to behave similarly towards CO_2 but do not deteriorate in the presence of oxygen. Based on the results of tests, the aqueous potassium salts (composed of sarcosine and proline) are considered to be the most promising solvents. The most common amino acids used in the gas treating solvents are glycine, alanine, dimethyl glycine, diethyl glycine, and a number of sterically hindered amino acids [65, 67, 68].

Results of many research groups showed that these solvents are suitable for application in membrane gas absorption units, because these solvents have better performance and degradation resistance than other chemical solvents [78]. Amino acid salts formed by neutralization of amino acids with an organic base such as amine showed better CO_2 absorption potential than amino acid salts from neutralization of amino acid salts from an inorganic base such as potassium hydroxide [79, 80]. Aronu et al. [69] studied the performance of amino acids neutralized with 3-(methylamino) propylamine (MAPA), glycine, β-alanine, and sarcosine. Their results indicated that sarcosine neutralized with MAPA has the best CO_2 absorption performance. Its performance is also enhanced by promoting with excess MAPA [69].

(3) *Ammonia.* Since ammonia is a toxic gas, prevention of ammonia "slip" to the atmosphere is a necessity. Despite this disadvantage, chilled ammonia process (CAP) was used for CO_2 separation (Figure 6). In the CAP, CO_2 is absorbed in an ammoniated solution at a lower absorption temperature (275–283 K) that reduced ammonia emissions from the CAP absorber. Ammonium carbonate solution resulted in approximately 38% carbon regeneration compared to MEA solution [70, 81, 82].

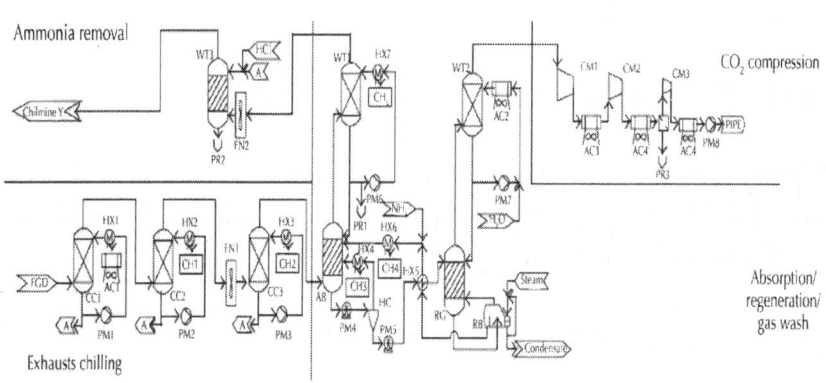

Figure 6: Schematic layout of CO_2 separation block based on the chilled ammonia process [92].

(4) *Aqueous Piperazine* (PZ). Piperazine (PZ) is as an additive used for amine systems to improve kinetics of CO_2 absorption, such as MDEA/PZ or MEA/PZ blends. Because PZ solubility in water is low, concentration of PZ is between 0.5 and 2.5 M. As indicated in Table 2, increasing the concentration of PZ in solution allows for increased solvent capacity and faster kinetic. The presence of

potassium in solution increases the concentration of CO_3^{2-} / HCO_3^- in solution; therefore, solution has buffering property. These competing effects yield a maximum fraction of reactive species at potassium to piperazine ratio of $2:1$ [75, 83, 84].

Adsorption

Adsorption operation can reduce energy and cost of the capture or separation of CO_2 in post-combustion capture. To achieve this goal, it is necessary to find adsorbents with suitable properties. In general, CO_2 adsorbent must have high selectivity and adsorption capacity and adequate adsorption/desorption kinetics, remain stable after several adsorption/desorption cycles, and possess good thermal and mechanical stability [51, 85–88]. The adsorbents used for CO_2 separation placed into two main categories: physical and chemical adsorbents.

Chemical Adsorption

Chemisorption is a subclass of adsorption, driven by a chemical reaction occurring at the exposed surface. Adsorption capacities of different chemical adsorbents are summarized in Table 3.

Table 3: Adsorption capacity of chemical adsorbents for post-combustion CO_2

Sorbent	Operating temperature (K)	Operating pressure (kPa)	CO_2 capture capacity (mol CO_2/kg sorbent)	Regeneration cycles,n	CO_2 capture capacity remained after n cycles (%)	Reference
Mesoporous (MgO)	298	101	1.8	3	100	[93]
CaO nanopods	873	101	17.5	50	61.1	[94]

CaO derived from nanosized $CaCO_3$	923	101	16.7	100	22.2	[93]
CaO-$MgAl_2O_4$(spinel nanoparticles)	923	101	9.1	65	84.6	[93]
Nano CaO/Al_2O_3	923	101	6.0	15	61.7	[93]
Lithium silicate nanoparticles	883	101	5.77	n.a.	n.a.	[93]
Nanocrystalline Li_2ZrO_3 particles	843	101	6.1	8	100	[93]
CaO/Al_2O_3	923	101	6.02	n.a.	n.a.	[93]
Lithium silicate	993	n.a.	8.18	n.a.	n.a.	[17]
Lithium zirconate	673	100	5.0	n.a.	n.a.	[93]
Lithium ortho-silicate	873	100	6.13	n.a.	n.a.	[93]
Calcium oxide	873	100	17.3	n.a.	n.a.	[93]
Magnesium hydroxide	473	1034	3.0	n.a.	n.a.	[93]
Mesoporous magnesium oxide	373	100	2.27	n.a.	n.a.	[93]
Lithium Silicate nano particles	873	101	5	n.a.	n.a.	[95]
HTl-HNa	573	134	1.109	50	93.3	[93]

A wide range of metals have been studied including [89]

- metal oxides: CaO, MgO,
- metal salts from alkali metal compounds: lithium silicate, lithium zirconate to alkaline earth metal compounds (i.e., magnesium oxide and calcium oxide),
- hydrotalcites and double salts.

In general, one mole of metal compound can react with one mole of CO_2 with a reversible reaction. The process consists of a series of cycles where metal oxides (such as CaO) at 923 K are transformed into metal carbonates form (such as $CaCO_3$) at 1123 K in a carbonation reactor to regenerate the sorbent and produce a concentrated stream of CO_2 suitable for storage [90, 91].

Considerable attention was paid to calcium oxide (CaO) as it has a high CO_2 adsorption capacity and high raw material availability (e.g., limestone) at a low cost. Lithium salts was recorded a good performance in CO_2 adsorption, but it gained less focus due to its high production cost. Although double salts can be easily regenerated due to low energy requirement, their stability has not been investigated [93, 96].

The reaction of CO_2 adsorption with Li_2ZrO_3 is reversible in the temperature range of 723–863 K. The capacity of lithium silicate (8.2 moL CO_2/kg sorbent at 993 K) is larger than that of lithium zirconate (4.85 moL/kg sorbent) [17].

Hydrotalcite (HT) contains layered structure with positively charged cations balanced by negatively charged anions [97, 98]. Adsorption and final capacity of different adsorption/desorption cycles are listed in Table 3.

One way for improving CO_2 adsorption efficiency is application of nanomaterials. Different nano-materials can be used for CO_2 separation (Table 3). However, nanomaterials always have high production cost with complicated synthesis process such as carbon nanotubes and graphite nanoplatelets [99, 100].

The main disadvantage of chemical adsorbents is difficult regeneration process, and application of these adsorbents needs more studies for finding new adsorbents [88, 95].

Physical Adsorption

Physisorption, also called physical adsorption, is a process in which the electronic structure of the atom or molecule is barely perturbed upon adsorption. If the CO_2 adsorption capacity of solid adsorbents reaches 3 mmoL/g, the required energy for adsorption will be less than 30–50% energy for absorption with optimum aqueous MEA [101]. The major physical adsorbents suggested for CO_2 adsorption include activated carbons and inorganic porous materials such as zeolites [102, 103]. The adsorption capacities of various physical adsorbents are summarized in Table 4.

Table 4: Adsorption capacity of physical adsorbents for post-combustion CO_2

Sorbent	Operating temperature (K)	Operating pressure (kPa)	CO_2 capture capacity (mol CO_2/kg sorbent)	Regeneration cycles,n	CO_2 capture capacity remained after n cycles (%)	Reference
Activated carbon	303	110	1.58	n.a.	n.a.	[93]
AC (4% KOH)	303	30	0.55	n.a.	n.a.	[93]
AC (EDA + EtOH)	303	30	0.53	n.a.	n.a.	[93]
AC (4% KOH + EDA + EtOH)	303	30	0.64	n.a.	n.a.	[45, 70,79]
NiO-ACs	298	101	2.227	n.a.	n.a.	[104]
13X	393	15.198	0.7	n.a.	n.a.	[105]
5A	393	15.198	0.38	n.a.	n.a.	[105, 106]
4A	393	15.198	0.5	n.a.	n.a.	[105]
WEG-592	393	15.198	0.6	n.a.	n.a.	[105]
APG-II	393	15.198	0.38	n.a.	n.a.	[105]
Na-Y	273	10.132	4.9	n.a.	n.a.	[105]
Na-X	373	101.32	1.24	2	n.a.	[105]
NaKA	373	101.32	3.88	—	n.a.	[105]
NaX-h	323	101.32	2.52	2	n.a.	[105]
NaX-h	373	101.32	1.37	2	n.a.	[105]
Na-X-c	323	101.32	2.14	2	n.a.	[105]
Na-X-c	373	101.32	1.41	2	n.a.	[105]
Cs-X-h	323	101.32	2.42	2	n.a.	[105]
Cs-X-h	373	101.32	1.48	2	n.a.	[105]
Cs-X-c	323	101.32	1.76	2	n.a.	[105]
Cs-X-c	373	101.32	1.15	n.a.	n.a.	[105]
MCM-41	298	100	0.62	n.a.	n.a.	[93]

MCM-41 (DEA)	348	100	1.26	n.a.	n.a.	[93]
MCM-41 (50% PEI)	348	100	2.52	n.a.	n.a.	[93]
Activated carbon	303	30	0.35	n.a.	n.a.	[93]
MCM-41 (50% PEI) "molecular basket"	348	100	2.95	n.a.	n.a.	[93]
PE-MCM-41	298	100	0.50	n.a.	n.a.	[93]
PE-MCM-41 (TRI)	298	100	2.85	n.a.	n.a.	[93]
PE-MCM-41 (DEA)	348	100	2.36	n.a.	n.a.	[93]
MCM-48	298	100	0.033	n.a.	n.a.	[93]
MCM-48 (APTS)	298	100	0.639	n.a.	n.a.	[93]
MCM-41	298	100	0.62	n.a.	n.a.	[93]
Molecular basket' MCM-41 (50% PEI)	348	100	2.5	8	96.0	[93]
PE-MCM-41 (TRI)	298	100	1.8	10	94.4	[93]
PE-MCM-41 (DEA)	298	100	2.9	7	96.6	[93]
MWNT	303	101	1.7	20	n.a.	[4, 93]
Unmodified [(Cu$_3$(btc)$_2$]*	298	1818	6.7	n.a.	n.a.	[101]
CNT@(Cu$_3$(btc)$_2$)	298	1818	13.52	n.a.	n.a.	[101]
MIL-101**	298	1010	0.84	n.a.	n.a.	[101]
MWCNT@MIL-101	298	1010	1.35	n.a.	n.a.	[101]
MOF-2	298	4545	3.20	n.a.	n.a.	[107]
MOF-177	298	4545	33.5	n.a.	n.a.	[107]
Zr-MOFs	273	988	8.1	n.a.	n.a.	[107]
Ca-Al LDH with ClO_4^-	406	1	3.55	n.a.	n.a.	[108]
Pd-GNP nanocomposite	298	1111	5.1	n.a.	n.a.	[109]

f-GNP	298	1111	4.3	n.a.	n.a.	[109]
Pd-GNP nano-composite	298	1111	4.5	n.a.	n.a.	[109]
f-GNP	298	1111	3.8	n.a.	n.a.	[109]
Pd-GNP nano-composite	298	1111	4.1	n.a.	n.a.	[109]
f-GNP	298	1111	3.3	n.a.	n.a.	[109]
Ceria-based oxides doped with 5% gallium (III)	298	101	0.282	n.a.	n.a.	[110]
Amine modified layered double hydroxides (LDHs)	298–353	101	0.74–1.75	n.a.	n.a.	[108]

*$Cu_3(btc)_2$; btc: 1,3,5-benzene-tricarboxylate.

**MIL-101 or $Cr_3(F,OH)(H_2O)_2O[(O_2C)C_6H_4(CO_2)]_3 \cdot nH_2O$ ($n \approx 25$) is one of the metal organic frameworks with Lewis acid sites that can be activated by removal of guest water molecules.

Coal is one of the adsorbents being suggested for CO_2 separation. The total amount of CO_2 that can be adsorbed in coal depends on its porosity, ash, and affinity for this molecule [111, 112]. Sakurovs et al. [113] showed that the ratio of maximum sorption capacity between CO_2 and methane decreases with increasing carbon content. The average CO_2/CH_4 sorption ratio is higher for moisture-equilibrated coal and decreases with increasing coal rank (1.4 for high rank coals to 2.2 for low rank coals) [114–116].

Activated carbon (AC) has a number of attractive characteristics, such as its high adsorption capacity, high hydrophobicity, low cost, and low energy requirement for regeneration [117–119]. Activated carbons are inexpensive, insensitive to moisture, and easy for regeneration. These adsorbents have well developed micro- and mesopore structures that are suitable for high CO_2 adsorption capacity at ambient pressure [120–122].

However, activated carbon CO_2/N_2 selectivities (ca. 10) are relatively low; zeolitic materials offer CO_2/N_2 selectivities 5–10 times greater than those of carbonaceous materials. The adsorption capacity and selectivity of zeolites are largely affected by their size, porous diameter, charge density, and chemical composition of cations in their porous structures. The average value of heat adsorption on zeolites (36 kJ/moL) is larger than for activated carbon (30 kJ/moL), confirming the mentioned affirmation. Moreover, activated carbon can be regenerated easily and completely. Also its capacity did not decay after 10 consecutive processes cycles [122–124].

Due to the increase in cost of raw materials, growing research interest has been focused on producing AC from agricultural wastes. Some of the agricultural wastes include the shells and stones of fruits, wastes resulting from the production of cereals, bagasse, and coir pith [100]. Rosas et al. [125] prepared hemp-derived AC monolith by phosphoric acid activation. The activated carbons from hemp stem are microporous materials and therefore suitable ones for hydrogen storage and CO_2 capture [126].

Siriwardane et al. [127] studied CO_2 adsorption on the molecular sieve 13X, 4A and activated carbon. The molecular sieve 13X showed better CO_2 separation than molecular sieve 4A. At lower pressures (<50 psi), activated carbon had a lower CO_2 separation than the molecular sieves, but adsorption was higher for activated carbon than molecular sieves at higher pressures [127, 128].

Liu et al. [129] indicated that zeolite 5A has higher volumetric capacities and less severe heat effect of the zeolite 13X. Chabazite zeolites were prepared and exchanged with alkali cations: Li, Na, K and alkaline-earth cations: Mg, Ca, Ba. Zhang et al. [130] studied the potential of these zeolites for CO_2 separation from flue gas by vacuum swing adsorption. It was found that NaCHA and CaCHA hold comparative advantages for high temperature CO_2 separation whilst NaX showed superior performance at relatively low temperatures [130]. In physical adsorption, the size and volume of the pores are important. Micropores are defined as pores, 2 nm in size, mesopores between 2 and 50 nm, and macropores, 50 nm in size. The micropores make better selective adsorption of CO_2 over CH_4 [131, 132].

Carbon nanotubes (CNTs) are the most famous among nano-hollow structured materials and their dimension ranges from 1 to 10 nm in diameter and from 200 to 500 nm in length [133]. Cinke et al. [134] indicated that purified single-walled carbon nanotubes (SWNTs) adsorbed CO_2 better than unpurified SWNT. In addition, multiwalled carbon nanotubes (MWNTs) showed stability for 20 cycles of adsorption and regeneration [135].

More recently, nanosystems researchers have synthesized and screened a large number of zeolitic-type materials known as zeolitic imidazolate frameworks (ZIFs). CO_2 capacities of the ZIFs are high, and selectivity against CO and N_2 is good [136, 137]. The results of researchers (Burchell and Judkins [138], Dave et al. [28], and Yong et al. [139]) indicated that the CO_2 adsorption efficiency of the honeycomb monolith is twice than activated carbon and 1.5 times greater than ZIF material [29]. Results of Kimber et al. [140] showed that CO_2 selectivity of honeycomb monolithic composite decreased with increasing in burn-off.

Graphite nanoplatelets (GNP) were prepared by acid intercalation followed by thermal exfoliation of natural graphite. Functionalized graphite nanoplatelets (f-GNP) were prepared by further treatment of GNP in acidic medium. Palladium (Pd) nanoparticles were decorated over f-GNP surface by chemical method [109, 141,142]. Adsorption capacity of this adsorbent is presented in Table 4.

The presence of several impurity gases (SO_x/NO_x/H_2O) greatly complicates the CO_2 separation processes. Therefore, conventional adsorption-based CO_2 separation processes rely on using a pretreatment stage to remove water, SO_x, and NO_x, which adds considerably to the overall cost. Also this prelayer can be used before the amine absorption column [143, 144]. Deng et al. [145] showed that the adsorption capacities follows the order $SO_2 > CO_2 > NO > N_2$ on both zeolites (5A and 13X). Comparing two different adsorbents, the better separation efficiency can be achieved by 5A zeolite [145].

Zhang et al. [130] focused on the effect of water vapour on the pressure/vacuum swing adsorption process. The selected adsorbents in this study were CDX (an alumina/zeolite blend), alumina, and 13X zeolite as these adsorbents are either the prelayer for water adsorption or the main CO_2 adsorption layer in the packed bed [130].

Metal-organic framework (MOF) materials are crystalline with two- or three-dimensional porous structures that can be synthesised with many of the functional capabilities of zeolites. Several MOFs have been proposed as adsorbents for CO_2 separation processes, and among these Cu-BTC [polymeric copper (II) benzene-1,3,5-tricarboxylate] has proved to be dedicated with CO_2 adsorption performances that are higher than those of typical adsorbents such as 13X zeolite [105, 107, 146, 147].

The MCM-41 material is one of the mesoporous products which was prepared by the hydrothermal method from mobil composition of matter (MCM) powders. Lu et al. [148] showed that mesoporous silica spherical particles (MSPs) can be synthesized using low-cost Na_2SiO_3 thus they can be cost-effective adsorbents for CO_2 separation from flue gas [149, 150].

Layered double hydroxides (LDHs) have general formula $M_{1-x}^{II}M_x^{III}(OH)_2\left[X_{x/c}^{g-}.nH_2O\right]$ with x typically in the range between 0.10 and 0.33. These materials can be readily and inexpensively synthesized with the desired characteristics for a particular application such as CO_2 adsorption [108, 151].

Adsorbent Modification

The role of CO_2 as a weak Lewis acid is well established. Because of the nature of CO_2, the surface of the physical adsorbents can be modified by adding basic groups, such as amine groups and metal oxides to improve CO_2 adsorption capacity or selectivity [152–154]. Three different methods for the production of these adsorbents were investigated: activation with CO_2, heat treatment with ammonia gas (amination and ammoxidation), and heat treatment with polyethylenimine (PEI). However, it has been suggested that amine modification can produce better and cheaper CO_2 adsorbents [24, 104, 155, 156].

Xu et al. [157, 158] designed selective "molecular basket" by grafting polyethylenimine (PEI) uniformly on MCM-41. CO_2 adsorption capacity of the adsorbent was 24 times higher than MCM-41 and 2 times higher than PEI [93]. The addition of ammonium hydroxide resulted in the Zr-MOF with a slight lower adsorption of CO_2 and CH_4; however, the selectivity of CO_2/CH_4 is significantly enhanced [159, 160]. Results of Abid et al. [107] showed that the selectivity of CO_2/CH_4 on Zr-MOF is between 2.2 and 3.8, while for Zr-MOF-NH_4selectivity is between 2.6 and 4.3.

A nitrogen-rich carbon with a hierarchical micro-mesopore structure exhibited a high CO_2 adsorption capacity (141 mg/g at 298 K, 1 atm), excellent separation efficiency (CO_2/N_2 selectivity is ca. 32), and excellent stability [161]. Plaza et al. [162] results showed that CO_2 adsorption capacity of the DETA-impregnated alumina (≥2.3 mmoL/g) exhibited is the highest.

Amine modified layered double hydroxides (LDHs) have been prepared by several different methods. Park et al. [163] used dodecyl sulfate (DS) intercalated LDH as precursor and added (3-aminopropyl) triethoxysilane (APTS) together with N-cetyl-N,N,N-trimethylammonium bromide (CTAB) [164]. The highest adsorption capacity of amine modified LDHs for CO_2 was achieved at 1.75 mmoL/g by MgAl N3 at 353 K and 1 bar. According to data in Table 4, this adsorbent has high CO_2 capacity at high temperature; therefore, this adsorbent is suitable for post-combustion CO_2 capture [108].

Wang et al. [114] reported that porous carbons with well-developed pore structures were directly prepared from a weak acid cation exchange resin (CER) by the carbonization of a mixture with Mg acetate in different ratios [108]. The main parameters of this adsorbent (such as CO_2 capacity) are indicated in Table 4.

Shafeeyan et al. [165] prepared different adsorbents based on the central composite design (CCD) with three independent variables (i.e., amination temperature, amination time, and the use of preheat treated (HTA) or preoxidized (OXA) sorbent as the starting material). They demonstrated that the optimum condition for obtaining an efficient CO_2 adsorbent is using a preoxidized sorbent and amination at 698 K for 2.1 h [165].

Table 4 compares CO_2 adsorption capacities and stability of different absorbents, which were studied for post-combustion CO_2 capture.

Different Cycles for CO₂ Adsorption

Five different regeneration strategies were demonstrated in a single-bed CO_2 adsorption unit: pressure swing adsorption (PSA), temperature swing adsorption (TSA), vacuum swing adsorption (VSA), electric swing adsorption (ESA), and a combination of vacuum and temperature swing adsorption (VTSA). The difference between these technologies is based on the strategy for regeneration of adsorbent after the adsorption step (Figure 7). In PSA applications, the pressure of the bed is reduced. VSA is preferred to the special PSA application where the desorption pressure is below atmospheric, whereas in TSA, the temperature is raised while pressure is maintained approximately constant, and in ESA the solid is heated by the Joule effect [166–169].

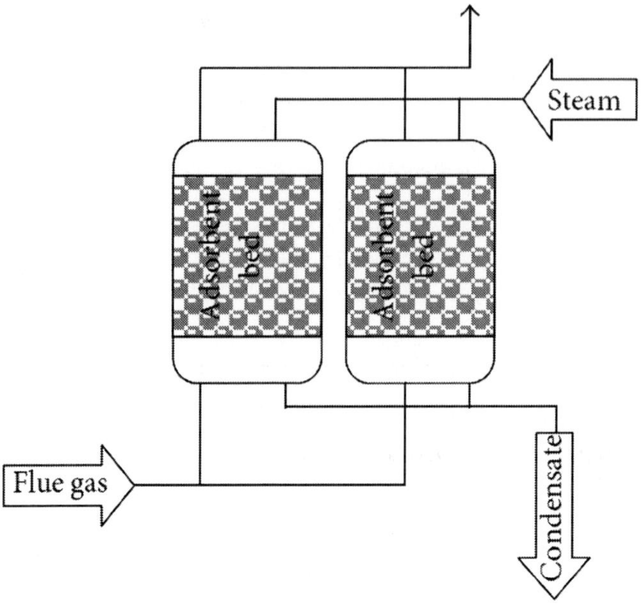
Steam
Adsorbent bed
Adsorbent bed
Flue gas
Condensate

(a)

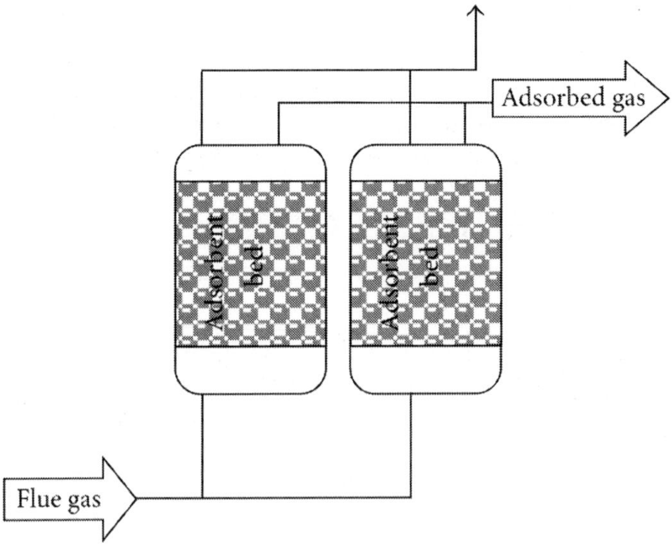
Adsorbed gas
Adsorbent bed
Adsorbent bed
Flue gas

(b)

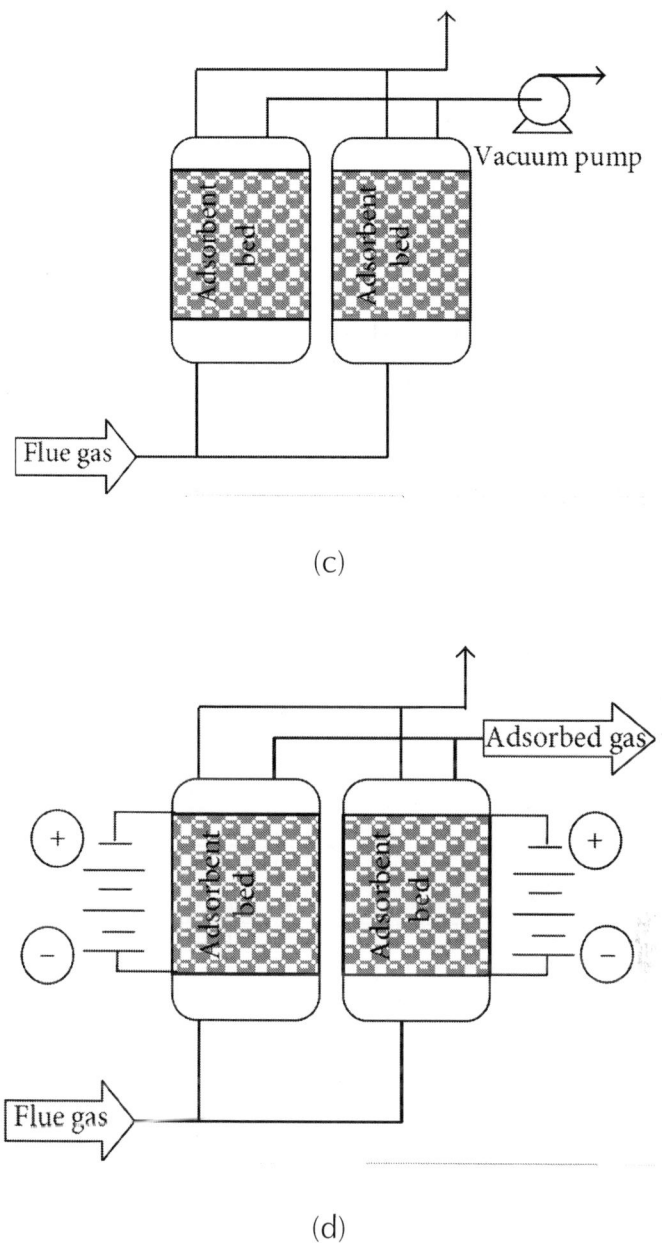

(c)

(d)

Figure 7: Schematic diagrams of various adsorption cycles, (a) TSA, (b) PSA, (c) VSA, and (d) ESA; thin lines indicated operation streams in re-generated step.

$$ESA < TSA < PSA < VSA < VTSA. \tag{1}$$

For the single-bed cycle configurations, the productivity and CO_2 recovery followed the sequence:

The performances of PSA, TSA, VSA, VTSA, and ESA processes for CO_2 separation are reported in Table 5. Since application of adsorption process for CO_2 capture in industrial scale is very important, in recent years some researches have been focused on this area; for example, Lucas et al. [170] studied the scale-up CO_2 adsorption with activated carbon.

Table 5: Comparison between several adsorption cycles for CO_2 separation process [166]

Process	CO_2 feed molar fraction (%) (other gases present)	CO_2 purity (%)	CO_2 recovery (%)
PSA	13 (O_2)	99.5	69
TSA	10	95	81
TSA	17	n.a.	40
ESA	10	23.33	92.57
VSA	15	90	90
VSA	17	n.a.	87
3-bed VSA	12	90–95	60–70
PSA/VSA	20	58–63	70–75
PSA/VSA	15 (H_2O)	59	87
VPSA	17	99.5–99.8	34–69
VPSA	16 (O_2)	99	53–70
PTSA	10	99	90
2-bed-2-step PSA	n.a.	18	90
VTSA	17	n.a.	97

Cryogenic Distillation

Cryogenic method utilized low temperatures for condensation, separation, and purification of CO_2 from flue gases (freezing point of pure CO_2 is 195.5 K at atmospheric pressure). Therefore, under the cryogenic separation process, the components can be separated by a series of compression, cooling, and expansion steps. It enables direct production of liquid CO_2 that can be stored or sequestered at high pressure via liquid pumping [171–173].

The advantages of this technology can be summarized as follows [6, 8, 174].

- Liquid CO_2 is directly produced, thus making it relatively easy to store or send for enhanced oil recovery.
- This technology is relatively straightforward, involving no solvents or other components.
- The cryogenic separation can be easy scaled-up to industrial-scale utilization.

The major disadvantages of this process are the large amount of energy required to provide the refrigeration and the CO_2 solidification under a low temperature, which causes several operational problems [176–178]. Therefore, more studies are required for reducing the cost of cryogenic separation.

Clodic et al. [179] indicated that the energy requirement for cryogenic process was in the range of 541–1119 kJ/kg CO_2. Zanganeh et al. [6] have constructed a pilot-scale CO_2 capture and compression unit (CO_2 CCU) that can separate CO_2 as liquid phase from the flue gas of oxy-fuel combustion. Their results showed that cryogenic is the most cost effective when the feed gas is available at high pressure. Therefore, cryogenic is not suitable for post-combustion and it is well effective for separation stream with high CO_2 concentration such as oxy-fuel combustion. Amann et al. [180] reported that conversion of O_2/CO_2 cycle was more efficient than amine scrubbing but more difficult to implement because of the specific gas turbine.

Xu et al. [175] studied a novel CO_2 cryogenic liquefaction and separation system (Figure 8). In this system, two-stage compression, two-stage refrigeration, two-stage separation, and sufficient recovery of cryogenic energy were adopted. The energy consumption for CO_2 recovery is only 0.395 MJ/kg CO_2. Furthermore, this CO_2 cryogenic separation system is more suitable for gas mixtures with high initial pressure and high CO_2 concentration [175].

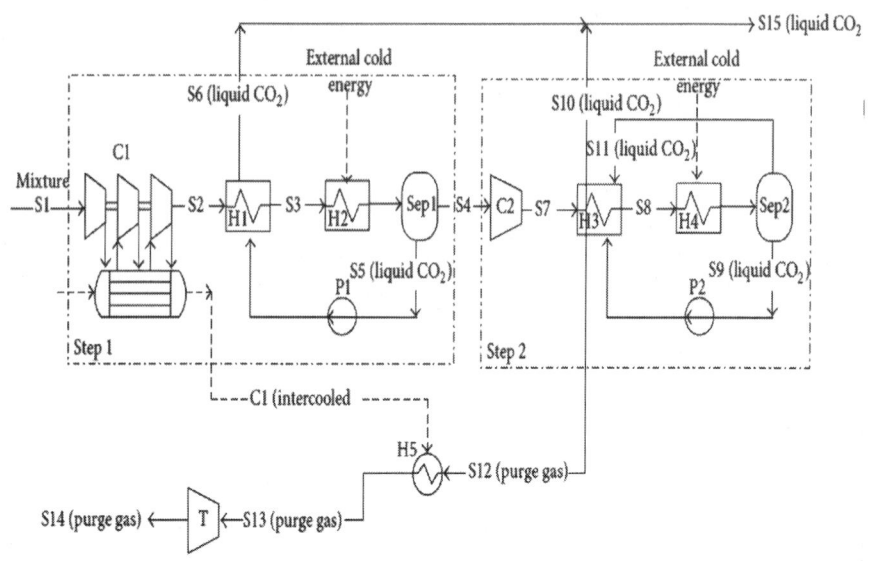

Figure 8: Novel CO_2 cryogenic liquefaction and separation system [175].

Song et al. [181] developed a novel cryogenic CO_2 capture system based on Stirling coolers (SC). The operation of Stirling cooler contains four processes: isothermal expansion, refrigeration under a constant volume, isothermal compression, and heating under a constant volume condition. This novel cryogenic system can condense and separate H_2O and CO_2 from flue gas. Their results showed that under the optimal temperature and flow rate, CO_2 recovery of the cryogenic process can reach 96% with 1.5 MJ/kg CO_2 energy consumption.

Tuinier et al. [182] exploited a novel cryogenic CO_2 capture process using dynamically operated packed beds (Figure 9). By the developed process, above 99% of CO_2 could be recovered from a flue gas containing 10 vol.% CO_2 and 1 vol.% H_2O with 1.8 MJ/kg CO_2 energy consumption [181].

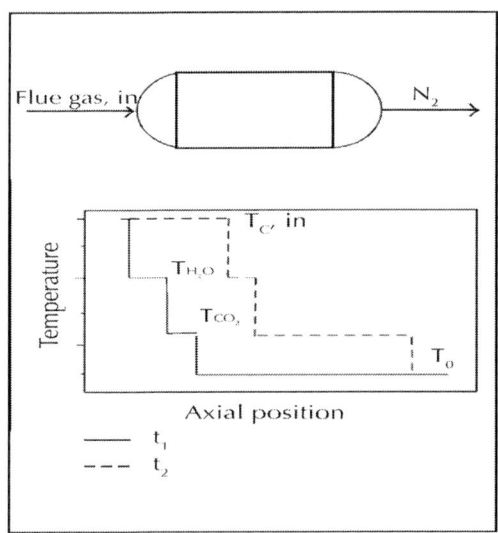

(a)

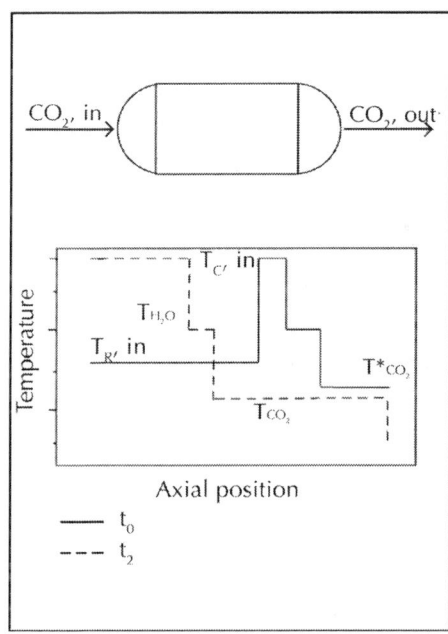

(b)

MCM-48	n.a.	n.a.	10200	n.a.	0.8	[189]
PEI-modified MCM-48	n.a.	363	14100	n.a.	80	[201]
Chitosan	1.75	295	100	n.a.	100	[192]
Swollen chitosan	1.5	383	482	n.a.	250	[192]
Arginine salt-chitosan	1.5	383	1500	n.a.	852	[194]
Polyacetylene						
Polytrimethyl-prop-1-ynyl-silane	n.a.	298	19000	1800	10.6	[193]
Poly-3,3-dimethyl-but-1-yne	n.a.	298	560	43	13.0	[193]
Poly-1-(dimethyl-trimethylsilanylmethyl-silanyl)-propyne	n.a.	298	310	21	14.8	[193]
Poly-1-[dimethyl-(2-trimethylsilanyl-ethyl)-silanyl]-propyne	n.a.	298	150	14	10.7	[193]
Polytrimethyl-(2-prop-1-ynyl-phenyl)-silane	n.a.	298	290	24	12.1	[193]
Poly-1-prop-1-ynyl-2-trifluoromethyl-benzene	n.a.	298	130	7.3	17.8	[193]
Poly-dec-2-yne	n.a.	298	130	14	9.3	[193]
Poly-1-chloro-dec-1-yne	n.a.	298	170	16	10.6	[193]
Poly-1-chloro-oct-1-yne	n.a.	298	130	11	11.8	[193]
Poly-1-chloro-hex-1-yne	n.a.	298	180	10	18	[193]

Polyhexyl-dimethyl-prop-1-ynyl-silane	n.a.	298	71	4.3	16.5	[193]
Polytrimethyl-(1-pentyl-prop-2-ynyl)-silane	n.a.	298	120	8.7	13.8	[193]
Polyhexyl-dimethyl-(1-pro-pyl-prop-2-ynyl)-silane	n.a.	298	70	6.3	11.1	[193]
Polyprop-1-ynyl-benzene	n.a.	298	25	2.2	11.4	[193]
Polybut-1-ynyl-benzene	n.a.	298	40	4.5	8.9	[193]
Polyoct-1-ynyl-benzene	n.a.	298	48	5.5	8.7	[193]
Polychloroethynyl-benzene	n.a.	298	23	1.0	23.0	[193]
Poly-1-ethynyl-2-methyl-benzene	n.a.	298	15	3.0	5.0	[193]
Polydimethyl-phenyl-(1-propyl-prop-2-ynyl)-silane	n.a.	298	54	2.5	21.6	[193]
Polyarylene ether						
6FPT-6FBPA	1.0	308	25.29	2.18	11.6	[193]
6FPT-BPA 1.0 35	1.0	308	18.53	1.37	13.5	[193]
6FPPy-6FBPA	1.0	308	29.46	2.39	12.32	[193]
6FPPy-BPA	1.0	308	21.44	1.70	12.6	[193]
Fixed site carrier membrane (FSCM)						
Polarix	2.0	303	10^7	n.a.	50	[202]
PAAM-PVA/PS	10	298	2.4×10^5	n.a.	80	[203]
PVAm/PVA blend	1.45	298	2.12×10^6	n.a.	145	[204]

PEI/PVA	n.a.	298	10^4	n.a.	230	[184]
PDMA/PS	2	296	3×10^5	n.a.	53	[143]
Polyamine						
PA12	10	308	120	n.a.	51	[152]
PA6	10	308	66	n.a.	56	[152]
Polyethyleneimine/polyvinyl butyral	0.132	318	380	n.a.	32	[193]
Poly[(2-N,N-dimethyl) aminoethyl methacrylate]	0.237	298	370	n.a.	111	[193]
Poly(vinylbenzyltrimethyl ammonium fluoride)	0.224	296	113	n.a.	983	[193]
Polyethyleneimine/ poly(vinyl alcohol)	0.355	298	650	n.a.	235	[193]
PEI/PDMS/PEBA1657/ PDMS	5	298	1.57×10^6	n.a.	64	[205]
Polyarylate						
BPA/IA	10	308	5.4	0.24	22.5	[193]
BPA/tBIA	10	308	24.2	1.20	20.2	[193]
HFBPA/IA	10	308	19.1	1.11	17.2	[193]
HFBPA/tBIA	10	308	56.9	3.88	14.7	[193]
PhTh/IA	10	308	6.74	0.28	24.1	[193]
PhTh/tBIA	10	308	23.8	1.09	21.8	[193]
FBP/IA	10	308	12.4	0.57	12.4	[193]

FBP/tBIA	10	308	36.8	1.93	19.1	[193]
TBBPA/IA	10	308	4.93	0.18	27.4	[193]
TBBPA/tBIA	10	308	21.5	0.90	23.9	[193]
TBHFBPA/IA	10	308	25.6	1.07	23.9	[193]
TBHFBPA/tBIA	10	308	85.1	4.47	19.0	[193]
TBPhTh/IA	10	308	8.34	0.29	28.8	[193]
TBPhTh/tBIA	10	308	30.6	1.28	23.9	[193]
TBFBP/IA	10	308	20.4	0.70	29.1	[193]
TBFBP/tBIA	10	308	69.5	2.94	23.6	[193]
DMBPA/IA	10	308	1.24	0.063	19.7	[193]
DMBPA/Tbia	10	308	8.0	0.39	20.5	[193]
TMBPA/IA	10	308	12.0	0.58	20.7	[193]
TMBPA/tBIA	10	308	44.6	2.52	17.7	[193]
DiisoBPA/IA	10	308	5.16	0.27	19.1	[193]
DiisoBPA/tBIA	10	308	16.1	1.08	14.9	[193]
DBDMBPA/IA	10	308	5.45	0.22	24.8	[193]
PhAnth/IA	10	308	9.0	0.36	25	[193]
PhAnth/tBIA	10	308	25.9	1.35	19.2	[193]
FBP/IA	10	308	12.4	0.57	21.8	[193]
FBP/tBIA	10	308	36.8	1.93	19.1	[193]
Polycarbonates						
PC	1–10	308	6.0–6.8	0.289–0.32	21	[193]

TMPC	1–10	308	17.58–18.6	1.0	18.6	[193]
TCPC	1	308	6.66	0.36	18.5	[193]
TBPC	1	308	4.23	0.182	23.2	[193]
HFPC	10	308	24	1.6	15.0	[193]
TMHFPC	10	308	111	7.4	15.0	[193]
NBPC	10	308	9.1	0.47	19.4	[193]
PCZ	10	308	2.2	0.105	21.0	[193]
PC-AP	2	308	9.48	0.361	26.3	[193]
FBPC	2	308	15.1	0.592	25.5	[193]
Polyethylene oxide						
PEO	7.8	298	8.1	0.07	140	[193]
PEO	4.4–14.6	308–318	13–52	0.24–1	55	[193]
PEO-PBT	n.a.	308	120	2	60	[193]
EO/EM/AGE (80/20/2)	n.a.	308	773	16.8	46	[193]
EO/EM/AGE (77/23/2.3)	n.a.	308	680	15.5	44	[193]
EO/EM/AGE (96/4/2.5)	n.a.	308	580	12.1	48	[193]
Polyimides						
Amine modified polyimide	0.368	308	186	n.a.	38	[193]
PMDA-BAPHF	6.8	308	11.8	0.66	17.8	[193]
PMDA-3BAPHF	6.8	308	6.12	0.29	21.1	[193]
PMDA-4,4'-ODA	6.8–10	308	1.14–2.7	0.049–0.1	23.3	[193]
PMDA-3,3'-ODA	6.8–10	308	0.50–3.55	0.018–0.145	24.5–27.8	[193]

PMDA-MDA	10	308	4.03	0.20	20.2	[193]
PMDA-IPDA	10	308	29.7	1.50	19.8	[193]
PMDA-BAPHF	10	308	17.6	0.943	18.7	[193]
PMDA-BATPHF	10	308	24.6	1.50	16.4	[193]
BPDA-BAHF	1–10	298–308	23–27.7	0.6–1.39	19.9–37.7	[193]
BPDA-mTrMPD	10	308	137	8.42	16.3	[193]
BTDA-4,4-ODA	10	308	0.625	0.0236	26.5	[193]
BTDA-BAPHF	10	308	4.37	0.195	22.4	[193]
BTDA-BAHF	10	308	10.1	0.45	22.4	[193]
BTDA-mTrMPD	10	308	30.9	1.55	19.9	[193]
BTDA-BAFL	1	298	15	0.39	38.5	[193]
PI	10	308	2.00	0.063	31.7	[193]
oMeCat-durene	1	303	27	0.83	33	[193]
mMeCat-durene	1	303	20	0.59	34	[193]
DMeCat-durene	1	303	63	2.05	31	[193]
mtBuCat-durene	1	303	71	2.55	28	[193]
oMeptBuCat-durene	1	303	67	2.5	27	[193]
TMeCat-durene	1	303	200	8.1	25	[193]
mMetCat-MDA	1	303	22	0.65	34	[193]
mtBuCat-MDA	1	303	63	2.2	29	[193]
TMeCat-MDA	1	303	110	3.8	30	[193]
TMeCat-TMB	1	303	39	1.2	33	[193]

DBuCat-TMB	1	303	95	4.9	19	[193]
mtBuCat-DMOB	1	303	6.7	0.21	32	[193]
TMeCat-6FiPDA	1	303	54	1.9	28	[193]
6F	3	n.a.	114	5.8	19.6	[193]
TMMPD	3	n.a.	600	35.1	17.1	[193]
IMDDM	3	n.a.	196	10.8	18.1	[193]
ODA	3	n.a.	25	0.97	25.8	[193]
Matrimid 5218	10	308	6.5	0.25	25.6	[193]
6FDA-based polyimides						
6FDA-pPDA	10	308	15.3	0.80	19.12	[193]
6FDA-pDiMPDA	10	303	42.7	2.67	16.0	[193]
6FDA-durene	10	308	440	35.60	12.4	[193]
6FDA-durene	10	303	456	35.50	12.85	[193]
6FDA-mPDA	6.8–10	308	8.23–9.20	0.36–0.447	20.6–22.7	[193]
6FDA-mMPDA	6.8–10	303	40.1–42.5	2.12–2.24	17.9–20.1	[193]
6FDA-mTrMPDA	10	308	431	31.6	13.6	[193]
6FDA-DATr	6.8	303	28.63	1.31	21.9	[193]
6FDA-DBTF	6.8	308	21.64	1.17	18.5	[193]
6FDA-PHDoeP	6.8	303	8.59	4.50	1.91	[193]
6FDA-PEPE	6.8	308	6.88	0.255	27.0	[193]
6FDA-PBEPE	6.8	303	2.50	0.099	25.3	[193]
6FDA-PMeaP	6.8	308	2.41	0.086	28.0	[193]

6FDA-3,4'ODA	10	303	6.11	0.259	23.6	[193]
6FDA-APAP	10	308	10.7	0.473	22.6	[193]
6FDA-pp'ODA	10	303	16.7	0.733	22.8	[193]
6FDA-BAPHF	10	308	19.1	0.981	19.5	[193]
6FDA-BATPHF	10	303	22.8	1.30	17.5	[193]
6FDA-BAHF	10	308	51.2	3.11	16.5	[193]
6FDA-1,5-NDA	10	308	23	1.1	21	[193]
6FDA-durene 24 h amidation	10	n.a.	11.6	1.33	8.75	[193]
6FDA-durene/mPDA (50/50)	10	n.a.	84.6	5.18	16.4	[193]
6FDA-durene/mPDA (50/50) 4 h amidation	10	n.a.	54.9	3.38	16.2	[193]
6FDA-durene/mPDA (50/50) 6 h amidation	10	n.a.	49.1	3.27	15.0	[193]
6FDA-durene/mPDA (50/50) 12 h amidation	10	n.a.	46.0	2.94	15.6	[193]
6FDA-durene/mPDA (50/50) 24 h amidation	10	n.a.	36.0	2.06	17.5	[193]
6FDA-durene/mPDA (50/50) 48 h amidation	10	n.a.	24.5	1.38	17.8	[193]
6FDA-FDA/HFBAPP (1/1)	1.1 kg/cm^2	303	465.0	19.9	23.4	[193]
6FDA-ODA	10	308	23	0.83	27.7	[193]

HMBIPSF	10	308	25.5	1.2	23.3	[193]
DMPSF-Z	10	308	1.4	0.057	24.6	[193]
PSF-AP	2	308	8.12	0.278	29.2	[193]
FBPSF	2	308	13.8	0.484	28.5	[193]
PSF-M	1	308	2.8	0.11	25.5	[193]
TMPSF-M	10	308	7.0	0.28	25.0	[193]
PSF-BPFL	1	308	10	0.25	40	[193]
3,4'-PSF	1	308	1.5	0.066	22.7	[193]
1,3-ADM PSF	35	308	7.2	0.33	21.8	[193]
2,2-ADM PSF	35	308	9.5	0.46	20.6	[193]
PSF (6% Br, 92% C≡CSiMe₃)	1	308	36.5	2.1	17.4	[193]
PSF (3% Br, 47% C≡CSiMe₃)	1	308	18.5	1.24	14.9	[193]
PSF (21% Br, 77% C≡CSiMe₃)	1	308	28.2	1.7	16.6	[193]
PSF (5% Br, 45% C≡CSiMe₃)	1	308	16.4	0.9	18.2	[193]
PSF	1	308	5.6	0.25	22.4	[193]
PSF-s-HBTMS	1	308	21	0.96	22.2	[193]
PSF-o-HBTMS	1	308	70	3.29	21.3	[193]
PSF-CH2-TMS	1	308	18	0.95	18.9	[193]
EM3	1	308	29	1.3	22	[193]

EM2	1	308	6.2	0.24	26	[193]
EM1	1	308	4.8	0.16	30	[193]
SM3 (degree of substitution = 2.0)	1	308	18	0.77	23	[193]
SM3 (degree of substitution = 1.0)	1	308	10	0.38	26	[193]
SM1	1	308	5.1	0.17	30	[193]
PPSF	1	308	3.2	0.10	32	[193]
RM3	1	308	27	1.9	14	[193]
RM2	1	308	6.7	0.60	11	[193]
RM1	1	308	6.9	0.61	11	[193]
HFPSF	1	308	12.0	0.67	17.9	[193]
HFPSF-o-HBTMS	1	308	105	5.63	18.6	[193]
HFPSF-s-TMS	1	308	41	2.0	20	[193]
HFPSF-o-TMS	1	308	84	4.7	18	[193]
HFPSF-TMS	1	308	110	6.3	18	[193]
TM6FPSF	1	308	72	4.0	18	[193]
TM6FPSF-s-TMS	1	308	96	5.2	19	[193]
TMPSF-TMS	1	308	32	1.51	21.3	[193]
TMPSF-s-TMS	1	308	66.3	3.07	21.6	[193]
TMPSF-HBTMS	1	308	72	3.36	21.4	[193]
Other membranes						

HQDPA-PDA	7	303	0.598	0.016	37.4	[193]
HQDPA-PDA	7	373	1.70	0.111	15.3	[193]
HQDPA-DBA	7	303	0.683	0.015	45.5	[193]
HQDPA-DBA	7	373	2.10	0.125	16.8	[193]
HQDPA-MDBA	7	303	1.18	0.034	34.7	[193]
HQDPA-MDBA	7	373	2.37	0.160	14.8	[193]
HQDPA-EDBA	7	303	2.26	0.077	29.4	[193]
HQDPA-EDBA	7	373	4.18	0.292	14.3	[193]
12H	5	308	4.6	0.21	21.9	[193]
6H6F	5	308	8.6	0.44	19.5	[193]
6F6H	5	308	8.9	0.42	21.2	[193]
12F	5	308	12.9	0.76	17.0	[193]
PBK	10	308	3.3	0.13	25.4	[193]
PBK-S	10	308	3.27	0.11	29.7	[193]
PBSF	10	308	10.8	0.47	23.0	[193]
PES/PI	4	308	1.15×10^5	n.a.	30	[193]
PPES	n.a.	273	0.92	0.027	34	[193]
PPESK	n.a.	273	0.75	0.042	18	[193]
20 percent DEA immobilized in 25.4 μm microporous polypropylene supports	0.16–1.67	298	974–4825	n.a.	56–276	[200]

Copolymers and polymer blend						
PEBA 2533 (hollow fiber)	6.8	273	260	n.a.	32	[206]
PEBA/PSF composite	3.4	273	6.1×10^5	n.a.	30	[206]
COPNA	n.a.	373	2990	n.a.	14	[200]
Pebax	n.a.	303	73	n.a.	15.6	[207]
Pebax/PEG10	n.a.	303	75	n.a.	15.8	[207]
Pebax/PEG20	n.a.	303	80	n.a.	15.9	[207]
Pebax/PEG30	n.a.	303	105	n.a.	15.1	[207]
Pebax/PEG40	n.a.	303	132	n.a.	15.1	[207]
Pebax/PEG50	n.a.	303	151	n.a.	15.5	[207]
Pebax/PEG-DME10	n.a.	303	123	n.a.	44	[208]
Pebax/PEG-DME20	n.a.	303	206	n.a.	45	[208]
Pebax/PEG-DME30	n.a.	303	300	n.a.	46	[208]
Pebax/PEG-DME40	n.a.	303	440	n.a.	42	[208]
Pebax/PEG-DME50	n.a.	303	606	n.a.	43	[208]
6FDA-TAB	10	308	54.0	2.8	19.3	[193]
6FDA/PMDA-TAB (50 : 50)	10	308	15.8	0.70	22.6	[193]
6FDA/PMDA-TAB (25 : 75)	10	308	3.13	0.098	31.9	[193]
6FDA/PMDA-TAB (10/90)	10	308	1.11	0.036	30.8	[193]
6FDA-TAB/DAM (75/25)	3	308	73.7	3.1	23.8	[193]
6FDA-TAB/DAM (50/50)	3	308	155	6.6	23.5	[193]
6FDA-DAM	3	308	370	29.5	12.5	[193]

6FDA/TMPDA	n.a.	308	400	23.5	17.02	[193]
6FDA/PMDA (1:6)-TM-MDA (CH$_2$Cl$_2$ cast)	10	308	187	11.7	16.0	[193]
6FDA/PMDA (1:6)-TM-MDA (NMP cast)	10	308	144	8.76	16.4	[193]
6FDA/PMDA (1:6)-TM-MDA (DMF cast)	10	308	88.6	5.16	17.2	[193]
MDI-BPA/PEG (75)	2	308	31	0.70	44	[193]
MDI-BPA/PEG (80)	2	308	48	1.0	47	[193]
MDI-BPA/PEG (85)	2	308	59	1.20	49	[193]
L/TDI (20)-BPA/PEG (90)	2	308	47	0.92	51	[193]
L/TDI (40)-BPA/PEG (85)	2	308	35	0.72	48	[193]
IPA-ODA/PEO3 (80)	2	308	58	1.1	53	[193]
BPDA-pp'ODA	n.a.	303	18000	n.a.	31	[155]
BPDA-ODA/DAT (oxi-dized)	n.a.	308	599	n.a.	40	[155]
BPDA-ODA/DABA/PEO1 (75)	2	308	2.7	0.048	56	[193]
BPDA-mDDS/PEO1 (80)	2	308	3.8	0.066	58	[193]
BPDA-ODA/DABA/PEO2 (70)	2	308	14	0.25	57	[193]
BPDA-ODA/DABA/PEO2 (80)	2	308	36	0.64	56	[193]
BPDA-ODA/PEO3 (75)	2	308	75	1.4	52	[193]

Membrane		Temp.				Ref.
BPDA-mDDS/PEO3 (75)	2	308	72	1.4	53	[193]
BPDA-mPD/PEO4 (80)	2	308	81	1.5	54	[193]
BPDA-ODA/PEO4 (80)	2	308	117	2.3	51	[193]
PMDA-ODA/DABA/PEO1 (80)	2	308	14	0.27	52	[193]
PMDA-ODA/PEO2 (75)	2	308	40	0.74	54	[193]
PMDA-mPD/PEO3 (80)	2	308	99	2.0	50	[193]
PMDA-APPS/PEO3 (80)	2	308	159	3.1	51	[193]
PMDA-APPS/PEO4 (70)	2	308	136	2.6	53	[193]
PMDA-mPD/PEO4 (80)	2	308	151	2.9	52	[193]
PMDA-ODA/PEO4 (80)	2	308	167	3.2	52	[152]
PMDA-pDDS/PEO4 (80)	2	308	238	4.9	49	[152]
PMDA/BTDA-BAFL (50:50)	1	298	43	1.3	33	[193]
PMDA/BTDA-BAFL (90:10)	1	298	130	3.8	34	[193]
BPDA-BAFL/HMDA (50:50)	1	298	0.54	0.014	39	[193]
PPES	n.a.	298	0.92	0.027	34	[193]
PPES/PPEK (3:1)	n.a.	298	2.94	0.074	40	[193]
PPES/PPEK (1:1)	n.a.	298	4.12	0.089	46	[193]
PPES/PPEK (1:3)	n.a.	298	2.06	0.026	39	[193]
PPES/PPEK (1:4)	n.a.	298	1.77	0.052	34	[193]

PPEK 18	n.a.	298	0.75	0.042	18	[193]
HQDPA-DPA/MDPA	7	303	0.957	0.023	41.2	[193]
HQDPA-DPA/MDPA	7	373	2.34	0.147	15.9	[193]
HQDPA-DPA/EDPA	7	303	1.334	0.036	37.6	[193]
HQDPA-DPA/EDPA	7	373	3.25	0.207	15.7	[193]
PI	10	308	2.00	0.063	31.7	[193]
PI/10PS	10	308	2.33	0.085	27.4	[193]
PI/15PS	10	308	2.32	0.09	25.8	[193]
PI/20PS	10	308	2.90	0.91	3.19	[193]
PI/25PS	10	308	4.29	0.91	4.71	[193]
PI/10PSVP	10	308	3.58	0.13	28.4	[193]
PI/15PSVP	10	308	3.71	0.14	26.5	[193]
PI/20PSVP	10	308	5.65	0.15	38.4	[193]
PI/25PSVP	10	308	6.55	1.55	4.31	[193]
NTDA-BDSA (30)/CARDO/ODA	3	303	70	1.7	41	[193]
NTDA-BDSA (30)/CARDO	3	303	164	4.5	36	[193]
NTDA-BDSA (30)/BAPHF	3	303	23	0.64	36	[193]
NTDA-BDSA (30)/ODA	3	303	5.2	0.1	52	[193]
6FDA-FDA/HFBAPP (1/1)	1.1 kg/cm²	303	465	19.9	23.4	[193]
6FDA-durene/pPDA (80/20)	10	308	230	16.88	13.62	[193]

6FDA-durene/pPDA (50/50)	10	308	126	7.74	16.28	[193]
6FDA-durene/pPDA (20/80)	10	308	59.26	2.81	21.09	[193]
6FDA-durene/3,3'-DDS (75/25)	10	308	84.7	5.91	14.3	[193]
6FDA-durene/3,3'-DDS (50/50)	10	308	19.8	1.09	18.2	[193]
6FDA-durene/3,3'-DDS (25/75)	10	308	5.12	0.26	19.7	[193]
6FDA-3,3'-DDS	10	308	1.84	0.08	22.7	[193]
6FDA-6FpDA-DABA-12.5	4	308	34.0	2.01	16.9	[193]
6FDA-6FpDA-DABA-12.5 annealed	4	308	70.8	4.50	15.7	[193]
6FDA-6FpDA-DABA-12.5 (22.5% TMOS)	4	308	30.9	1.70	18.2	[193]
6FDA-6FpDA-DABA-12.5 (22.5% TMOS) annealed	4	308	47.6	3.16	15.1	[193]
6FDA-6FpDA-DABA-12.5 (15.0% MTMOS)	4	308	44.0	2.53	17.4	[193]
6FDA-6FpDA-DABA-12.5 (15.0% MTMOS) annealed	4	308	110	7.07	15.6	[193]
6FDA-6FpDA-DABA-12.5 (15.0% PTMOS) 4 35	4	308	32.3	1.80	17.9	[193]

6FDA-6FpDA-DABA-12.5 (15.0% PTMOS) annealed	4	308	91.8	5.59	16.4	[193]
6FDA-6FpDA-DABA-12.5 (22.5% PTMOS)	4	308	30.7	1.88	16.3	[193]
6FDA-6FpDA-DABA-12.5 (22.5% PTMOS) annealed	4	308	90.9	5.87	15.5	[193]
6FDA-6FpDA-DABA-25	4	308	20.3	1.20	16.9	[193]
6FDA-6FpDA-DABA-25 annealed	4	308	77.3	4.85	15.9	[193]
6FDA-6FpDA-DABA-25 (22.5% TMOS)	4	308	15.7	1.06	14.8	[193]
6FDA-6FpDA-DABA-25 (22.5% TMOS) annealed	4	308	79.8	4.87	16.4	[193]
6FDA-6FpDA-DABA-25 (15.0% MTMOS)	4	308	16.6	1.07	15.5	[193]
6FDA-6FpDA-DABA-25 (15.0% MTMOS) annealed	4	308	81.1	5.07	16.0	[193]
6FDA-6FpDA-DABA-25 (22.5% MTMOS)	4	308	16.6	1.07	15.5	[193]
6FDA-6FpDA-DABA-25 (22.5% MTMOS) annealed	4	308	60.1	3.837	15.7	[193]
6FDA-6FpDA-DABA-25 (15.0% PTMOS)	4	308	18.4	0.94	19.6	[193]
6FDA-6FpDA-DABA-25 (15.0% PTMOS) annealed	4	308	104	6.25	16.6	[193]

6FDA-6FpDA-DABA-25 (22.5% PTMOS)	4	308	19.1	0.98	19.5	[193]
6FDA-6FpDA-DABA-25 (22.5% PTMOS) annealed	4	308	104	6.25	16.6	[193]
Poly(5 : 5 BPA/BN)	5	308	5.71	0.19	30.1	[193]
Poly(7 : 3 BPA/BN)	5	308	4.62	0.16	28.9	[193]
Cross-linking polymers						
Poly(ethylene oxide-co-epichlorohydrin) (1 : 1) 1.1%	300	298	15.0	2.3	6.52	[193]
Poly(ethylene oxide-co-epichlorohydrin) (1 : 1) 2%	300	298	14.9	1.0	14.9	[193]
Poly(ethylene oxide-co-epichlorohydrin) (1 : 1) 5%	300	298	16.1	0.5	32.2	[193]
DM14/MM9 (100/0)	0.967	298	45	0.66	68	[193]
DM14/MM9 (100/0)	0.967	323	107	2.8	38	[193]
DM14/MM9 (90/10)	0.967	298	62	0.90	69	[193]
DM14/MM9 (90/10)	0.967	323	133	3.4	39	[193]
DM14/MM9 (70/30)	0.967	298	96	1.5	66	[193]
DM14/MM9 (70/30)	0.967	323	195	5.4	36	[193]
DM14/MM9 (50/50)	0.967	298	144	2.25	64	[193]
DM14/MM9 (50/50)	0.967	323	260	7.2	36	[193]
DM14/MM9 (30/70)	0.967	298	210	3.3	63	[193]
DM14/MM9 (30/70)	0.967	323	350	10.6	33	[193]

capture technologies and opportunities in Canada, CANMET Energy Technology Centre Natural Resources Canada, 2003.

48.   H. J. Herzog, "The economics of CO_2 separation and capture," Journal of the Franklin Institute, vol. 7, pp. 13–24, 2000.

49.   G. Pellegrini, R. Strube, and G. Manfrida, "Comparative study of chemical absorbents in postcombustion CO_2 capture," Energy, vol. 35, no. 2, pp. 851–857, 2010.

50.   N. MacDowell, N. Florin, A. Buchard et al., "An overview of CO_2 capture technologies," Energy and Environmental Science, vol. 3, no. 11, pp. 1645–1669, 2010. · ·

51.   X. P. Li Gang, A. Webley Paul, Zhang Jun, and R. Singh, "Competition of CO_2/H_2O in adsorption based CO_2 capture," Energy Procedia, vol. 1, pp. 1123–1130.

52.   P. Singh, J. P. M. Niederer, and G. F. Versteeg, "Structure and activity relationships for amine based CO_2absorbents-I," International Journal of Greenhouse Gas Control, vol. 1, no. 1, pp. 5–10, 2007.

53.   S. Ma›mun, "Selection and characterization of new absorbents for carbon dioxide capture," in Chemical Engineering, Faculty of Natural Science and Technology, 2005.

54.   S. Cavenati, C. A. Grande, and A. E. Rodrigues, "Removal of carbon dioxide from natural gas by vacuum pressure swing adsorption," Energy and Fuels, vol. 20, no. 6, pp. 2648–2659, 2006.

55.   J. David, Economic evaluation of leading technology options 23 for sequestration of carbon dioxide [M.S. thesis], Chemical Engineering Practice Massachusetts Institute of Technology, 2000.

56.   D. L. Albritton, T. Barker, I. A. Bashmakov et al., Climate Change 2001: Synthesis Report, edited by D. J. Dokken, M. Noguer, ,P. V. d Linden,C. Johnson, J. Pan, Cambridge University Press, 2001.

57.   M. Wang, A. Lawal, P. Stephenson, J. Sidders, and C. Ramshaw, "Post-combustion CO_2 capture with chemical absorption: a

state-of-the-art review," Chemical Engineering Research and Design, vol. 89, no. 9, pp. 1609–1624, 2011.

58. J. Gabrielsen, H. F. Svendsen, M. L. Michelsen, E. H. Stenby, and G. M. Kontogeorgis, "Experimental validation of a rate-based model for CO_2 capture using an AMP solution," Chemical Engineering Science, vol. 62, no. 9, pp. 2397–2413, 2007.

59. R. Idem, M. Wilson, P. Tontiwachwuthikul et al., "Pilot plant studies of the CO_2 capture performance of aqueous MEA and mixed MEA/MDEA solvents at the University of Regina CO_2 capture technology development plant and the boundary dam CO_2 capture demonstration plant," Industrial and Engineering Chemistry Research, vol. 45, no. 8, pp. 2414–2420, 2006.

60. M. Lucquiaud and J. Gibbins, "On the integration of CO_2 capture with coal-fired power plants: a methodology to assess and optimise solvent-based post-combustion capture systems," Chemical Engineering Research and Design, vol. 89, no. 9, pp. 1553–1571, 2011.

61. J. N. Knudsen, J. N. Jensen, P. J. Vilhelmsen, and O. Biede, "Experience with CO_2 capture from coal flue gas in pilot-scale: testing of different amine solvents," Energy Procedia, vol. 1, no. 1, pp. 783–790, 2009.

62. P. H. M. Feron, "Exploring the potential for improvement of the energy performance of coal fired power plants with post-combustion capture of carbon dioxide," International Journal of Greenhouse Gas Control, vol. 4, no. 2, pp. 152–160, 2010.

63. F. Qin, S. Wang, A. Hartono, H. F. Svendsen, and C. Chen, "Kinetics of CO_2 absorption in aqueous ammonia solution," International Journal of Greenhouse Gas Control, vol. 4, no. 5, pp. 729–738, 2010.

64. H. P. Mangalapally, R. Notz, S. Hoch et al., "Pilot plant experimental studies of post combustion CO_2 capture by reactive absorption with MEA and new solvents," Energy Procedia, vol. 1, pp. 963–970, 2009.

65. P. S. Kumar, J. A. Hogendoorn, G. F. Versteeg, and P. H. M. Feron, "Kinetics of the reaction of CO_2 with aqueous potassium salt of taurine and glycine," AIChE Journal, vol. 49, no. 1, pp. 203–213, 2003.

66. S. A. Freeman, R. Dugas, D. van Wagener, T. Nguyen, and G. T. Rochelle, "Carbon dioxide capture with concentrated, aqueous piperazine," Energy Procedia, vol. 1, pp. 1489–1496, 2009.

67. J. V. Holst, G. F. Versteeg, D. W. F. Brilman, and J. A. Hogendoorn, "Kinetic study of CO_2 with various amino acid salts in aqueous solution," Chemical Engineering Science, vol. 64, no. 1, pp. 59–68, 2009.

68. E. S. Hamborg, J. P. M. Niederer, and G. F. Versteeg, "Dissociation constants and thermodynamic properties of amino acids used in CO_2 absorption from (293 to 353) K," Journal of Chemical and Engineering Data, vol. 52, no. 6, pp. 2491–2502, 2007.

69. U. E. Aronu, H. F. Svendsen, and K. A. Hoff, "Investigation of amine amino acid salts for carbon dioxide absorption," International Journal of Greenhouse Gas Control, vol. 4, no. 5, pp. 771–775, 2010.

70. J. T. Yeh, K. P. Resnik, K. Rygle, and H. W. Pennline, "Semi-batch absorption and regeneration studies for CO_2 capture by aqueous ammonia," Fuel Processing Technology, vol. 86, no. 14-15, pp. 1533–1546, 2005.

71. C. H. Yu, C. H. Huang, and C. S. Tan, "A Review of CO_2 Capture by Absorption and Adsorption,"Aerosol and Air Quality Research, vol. 12, pp. 745–769, 2012.

72. B. E. Gurkan, C. Juan, E. M. Mindrup et al., "Chemically complexing ionic liquids for post-combustion CO_2 capture," in Clearwater Clean Coal Conference, pp. 6–10, Clearwater, Fla, USA, 2010.

73. E. D. Bates, R. D. Mayton, I. Ntai, and J. H. Davis Jr., "CO_2 capture by a task-specific ionic liquid,"Journal of the American Chemical Society, vol. 124, no. 6, pp. 926–927, 2002.

74. S. Baj, A. Siewniak, A. Chrobok, T. Krawczyk, and A. Sobolewski, "Monoethanolamine and ionic liquid aqueous solutions as effective systems for CO_2 capture," Journal of Chemical Technology and Biotechnology, vol. 88, pp. 1220–1227, 2012.

75. J. P. Ciferno, D. Lang, and G. T. Rochelle, Carbon Dioxide Capture by Absorption with Potassium Carbonate, University of Texas, 2010.

76. J. T. Cullinane and G. T. Rochelle, "Thermodynamics of aqueous potassium carbonate, piperazine, and carbon dioxide," Fluid Phase Equilibria, vol. 227, no. 2, pp. 197–213, 2005.

77. H. P. Mangalapally and H. Hasse, "Pilot plant experiments with mea and new solvents for post combustion CO_2 capture by reactive absorption," Energy Procedia, vol. 4, pp. 1–8, 2011.

78. J. Brouwer, P. Feron, and N. Ten Asbroek, "Amino-acid salts for CO_2 capture from flue gases," inProceedings of the 4th Annual Conference on Carbon Capture & Sequestration, 2009.

79. D. Kang, S. Park, H. Jo, J. Min, and J. Park, "Solubility of CO_2 in amino-acid-based solutions of (potassium sarcosinate), (potassium alaninate + piperazine), and (potassium serinate + piperazine),"Journal of Chemical & Engineering Data, vol. 58, pp. 1787–1791, 2013.

80. B. Farid and E. Fadwa, "Front matter," in Proceedings of the 2nd Annual Gas Processing Symposium, p. 488, Elsevier, Doha, Qatar, 2010.

81. R. M. Davidson, Post-Combustion Carbon Capture from Coal Fired Plants: Solvent Scrubbing, IEA Clean Coal Centre, 2007.

82. V. Darde, K. Thomsen, W. J. van Well, and E. H. Stenby, "Chilled ammonia process for CO_2 capture,"Energy Procedia, vol. 1, pp. 1035–1042, 2009.

83. S. Bishnoi and G. T. Rochelle, "Thermodynamics of piperazine/methyldiethanolamine/water/carbon dioxide," Industrial and

Engineering Chemistry Research, vol. 41, no. 3, pp. 604–612, 2002. ·

84. A. Bajpai and M. K. Mondal, "Equilibrium solubility of CO_2 in aqueous mixtures of DEA and AEEA,"Journal of Chemical & Engineering Data, vol. 58, pp. 1490–1495, 2013.

85. A. L. Chaffee, G. P. Knowles, Z. Liang, J. Zhang, P. Xiao, and P. A. Webley, "CO_2 capture by adsorption: materials and process development," International Journal of Greenhouse Gas Control, vol. 1, no. 1, pp. 11–18, 2007.

86. J.-R. Li, Y. Ma, M. C. McCarthy et al., "Carbon dioxide capture-related gas adsorption and separation in metal-organic frameworks," Coordination Chemistry Reviews, vol. 255, no. 15-16, pp. 1791–1823, 2011.

87. L.-Y. Meng and S.-J. Park, "Influence of MgO template on carbon dioxide adsorption of cation exchange resin-based nanoporous carbon," Journal of Colloid and Interface Science, vol. 366, no. 1, pp. 135–140, 2012.

88. M. Sevilla and A. B. Fuertes, "CO_2 adsorption by activated templated carbons," Journal of Colloid and Interface Science, vol. 366, no. 1, pp. 147–154, 2012.

89. M. Martunus, Z. Helwani, A. D. Wiheeb, J. Kim, and M. R. Othman, "Improved carbon dioxide capture using metal reinforced hydrotalcite under wet conditions," International Journal of Greenhouse Gas Control, vol. 7, pp. 127–136, 2012.

90. B. Dou, Y. Song, Y. Liu, and C. Feng, "High temperature CO_2 capture using calcium oxide sorbent in a fixed-bed reactor," Journal of Hazardous Materials, vol. 183, no. 1–3, pp. 759–765, 2010.

91. M. Kotyczka-moranska, G. Tomaszewicz, and G. Labojko, "Comparison of different methods for enhancing CO_2 capture by CaO-based sorbents. Review," Physicochemical Problems of Mineral Processing, vol. 48, pp. 77–90, 2012.

92. G. Valenti, D. Bonalumi, and E. Macchi, "A parametric investigation of the chilled ammonia process from energy and economic perspectives," Fuel, vol. 101, pp. 74–83, 2011.

93. Z. H. Lee, K. T. Lee, S. Bhatia, and A. R. Mohamed, "Post-combustion carbon dioxide capture: evolution towards utilization of nanomaterials," Renewable and Sustainable Energy Reviews, vol. 16, no. 5, pp. 2599–2609, 2012.

94. Z. Xiang, Z. Hu, D. Cao et al., "Metal-organic frameworks with incorporated carbon nanotubes: improving carbon dioxide and methane storage capacities by lithium doping," Angewandte Chemie, vol. 50, no. 2, pp. 491–494, 2011.

95. K. Essaki, M. Kato, and K. Nakagawa, "CO_2 removal at high temperature using packed bed of lithium silicate pellets," Journal of the Ceramic Society of Japan, vol. 114, no. 1333, pp. 739–742, 2006.

96. C. S. Martavaltzi and A. A. Lemonidou, "Development of new CaO based sorbent materials for CO_2 removal at high temperature," Microporous and Mesoporous Materials, vol. 110, no. 1, pp. 119–127, 2008.

97. R. Besson, M. Rocha Vargas, and L. Favergeon, "CO_2 adsorption on calcium oxide: an atomic-scale simulation study," Surface Science, vol. 606, no. 3-4, pp. 490–495, 2012.

98. S. Miyata, "Anion-exchange properties of hydrotalcite-like compounds," Clays & Clay Minerals, vol. 31, no. 4, pp. 305–311, 1983. ·

99. A. R. Mohamed, S. Bhatia, K. T. Lee, C. Y. H. Foo, Z. H. Lee, and N. A. Razali, "Nanomaterials as environmentally compatible next generation green carbon capture and utilization materials,"Transactions on GIGAKU, vol. 1, Article ID 01006, pp. 1–7, 2012.

100. M. Songolzadeh, M. Takht Ravanchi, and M. Soleimani, "Carbon dioxide capture and storage: a general review on adsorbents," World Academy of Science, Engineering and Technology, vol. 70, pp. 225–232, 2012.

101. M. Anbia and V. Hoseini, "Development of MWCNT@MIL-101 hybrid composite with enhanced adsorption capacity for carbon dioxide," Chemical Engineering Journal, vol. 191, pp. 326–330, 2012.

102. L.-Y. Lin and H. Bai, "Continuous generation of mesoporous silica particles via the use of sodium metasilicate precursor and their potential for CO_2 capture," Microporous and Mesoporous Materials, vol. 136, no. 1–3, pp. 25–32, 2010.

103. D. M. D›Alessandro, B. Smit, and J. R. Long, "Carbon dioxide capture: prospects for new materials,"Angewandte Chemie, vol. 49, no. 35, pp. 6058–6082, 2010. · ·

104. D. I. Jang and S. J. Park, "Influence of nickel oxide on carbon dioxide adsorption behaviors of activated carbons," Fuel, vol. 102, pp. 439–444, 2012.

105. S. Choi, J. H. Drese, and C. W. Jones, "Adsorbent materials for carbon dioxide capture from large anthropogenic point sources," ChemSusChem, vol. 2, no. 9, pp. 796–854, 2009.

106. J. A. Delgado, M. A. Uguina, J. L. Sotelo, and B. Ruíz, "Fixed-bed adsorption of carbon dioxide-helium, nitrogen-helium and carbon dioxide-nitrogen mixtures onto silicalite pellets," Separation and Purification Technology, vol. 49, no. 1, pp. 91–100, 2006. · ·

107. H. R. Abid, G. H. Pham, H.-M. Ang, M. O. Tade, and S. Wang, "Adsorption of CH_4 and CO_2 on Zr-metal organic frameworks," Journal of Colloid and Interface Science, vol. 366, no. 1, pp. 120–124, 2012.

108. J. Wang, L. A. Stevens, T. C. Drage, and J. Wood, "Preparation and CO_2 adsorption of amine modified Mg-Al LDH via exfoliation route," Chemical Engineering Science, vol. 68, no. 1, pp. 424–431, 2012.

109. A. K. Mishra and S. Ramaprabhu, "Palladium nanoparticles decorated graphite nanoplatelets for room temperature carbon dioxide adsorption," Chemical Engineering Journal, vol. 187, pp. 10–15, 2012.

110. G. Finos, S. Collins, G. Blanco et al., "Infrared spectroscopic study of carbon dioxide adsorption on the surface of cerium-gallium mixed oxides," Catalysis Today, vol. 180, no. 1, pp. 9–18, 2012.

111. R. P. Grimm, K. A. Eriksson, N. Ripepi, C. Eble, and S. F. Greb, "Seal evaluation and confinement screening criteria for beneficial carbon dioxide storage with enhanced coal bed methane recovery in the Pocahontas Basin, Virginia," International Journal of Coal Geology, vol. 90-91, pp. 110–125, 2012.

112. B. Guo, L. Chang, and K. Xie, "Adsorption of carbon dioxide on activated carbon," Journal of Natural Gas Chemistry, vol. 15, no. 3, pp. 223–229, 2006.

113. R. Sakurovs, S. Day, and S. Weir, "Relationships between the sorption behaviour of methane, carbon dioxide, nitrogen and ethane on coals," Fuel, vol. 97, pp. 725–729, 2012.

114. P. Weniger, J. Francǔ, P. Hemza, and B. M. Krooss, "Investigations on the methane and carbon dioxide sorption capacity of coals from the SW Upper Silesian Coal Basin, Czech Republic," International Journal of Coal Geology, vol. 93, pp. 23–39, 2012.

115. C. Garnier, G. Finqueneisel, T. Zimny et al., "Selection of coals of different maturities for CO_2 Storage by modelling of CH_4 and CO_2 adsorption isotherms," International Journal of Coal Geology, vol. 87, no. 2, pp. 80–86, 2011.

116. J. C. Abanades, E. S. Rubin, and E. J. Anthony, "Sorbent cost and performance in CO_2 capture systems,"Industrial and Engineering Chemistry Research, vol. 43, no. 13, pp. 3462–3466, 2004. ·

117. T. C. Drage, J. M. Blackman, C. Pevida, and C. E. Snape, "Evaluation of activated carbon adsorbents for CO_2 capture in gasification," Energy and Fuels, vol. 23, no. 5, pp. 2790–2796, 2009.

118. W. Shen, S. Zhang, Y. He, J. Li, and W. Fan, "Hierarchical porous polyacrylonitrile-based activated carbon fibers for

135. A. F. Portugal, P. W. J. Derks, G. F. Versteeg, F. D. Magalhães, and A. Mendes, "Characterization of potassium glycinate for carbon dioxide absorption purposes," Chemical Engineering Science, vol. 62, no. 23, pp. 6534–6547, 2007.

136. R. Banerjee, A. Phan, B. Wang et al., "High-throughput synthesis of zeolitic imidazolate frameworks and application to CO_2 capture," Science, vol. 319, no. 5865, pp. 939–943, 2008.

137. K. S. Park, Z. Ni, A. P. Côté et al., "Exceptional chemical and thermal stability of zeolitic imidazolate frameworks," Proceedings of the National Academy of Sciences of the United States of America, vol. 103, no. 27, pp. 10186–10191, 2006.

138. T. D. Burchell and R. R. Judkins, "Passive CO_2 removal using a carbon fiber composite molecular sieve,"Energy Conversion and Management, vol. 37, no. 6–8, pp. 947–954, 1996. ·

139. Z. Yong, V. Mata, and A. E. Rodrigues, "Adsorption of carbon dioxide at high temperature—a review,"Separation and Purification Technology, vol. 26, no. 2-3, pp. 195–205, 2002.

140. G. M. Kimber, M. Jagtoyen, Y. Q. Fei, and F. J. Derbyshire, "Fabrication of carbon fibre composites for gas separation," Gas Separation and Purification, vol. 10, no. 2, pp. 131–136, 1996. ·

141. L. M. Viculis, J. J. Mack, O. M. Mayer, H. T. Hahn, and R. B. Kaner, "Intercalation and exfoliation routes to graphite nanoplatelets," Journal of Materials Chemistry, vol. 15, no. 9, pp. 974–978, 2005.

142. A. K. Mishra and S. Ramaprabhu, "Study of CO_2 adsorption in low cost graphite nanoplatelets,"International Journal of Chemical Engineering and Applications, vol. 1, pp. 266–269, 2010.

143. R. Du, X. Feng, and A. Chakma, "Poly(N,N-dimethylaminoethyl methacrylate)/polysulfone composite membranes for gas separations," Journal of Membrane Science, vol. 279, no. 1-2, pp. 76–85, 2006.

144. K. Kumar, C. N. Dasgupta, B. Nayak, P. Lindblad, and D. Das, "Development of suitable photobioreactors for CO_2 sequestration addressing global warming using green algae and cyanobacteria," Bioresource Technology, vol. 102, no. 8, pp. 4945–4953, 2011.

145. H. Deng, H. Yi, X. Tang, Q. Yu, P. Ning, and L. Yang, "Adsorption equilibrium for sulfur dioxide, nitric oxide, carbon dioxide, nitrogen on 13X and 5A zeolites," Chemical Engineering Journal, vol. 188, pp. 77–85, 2012.

146. N. Gargiulo, F. Pepe, and D. Caputo, "Modeling carbon dioxide adsorption on polyethylenimine-functionalized TUD-1 mesoporous silica," Journal of Colloid and Interface Science, vol. 367, no. 1, pp. 348–354, 2012.

147. A. R. Millward and O. M. Yaghi, "Metal-organic frameworks with exceptionally high capacity for storage of carbon dioxide at room temperature," Journal of the American Chemical Society, vol. 127, no. 51, pp. 17998–17999, 2005.

148. C. Lu, H. Bai, F. Su, W. Chen, J. F. Hwang, and H.-H. Lee, "Adsorption of carbon dioxide from gas streams via mesoporous spherical-silica particles," Journal of the Air and Waste Management Association, vol. 60, no. 4, pp. 489–496, 2010.

149. A. Boonpoke, S. Chiarakorn, N. Laosiripojana, S. Towprayoon, and A. Chidthaisong, "synthesis of activated carbon and MCM-41 from bagasse and rice husk and their carbon dioxide adsorption capacity," Journal of Sustainable Energy & Environmentn, vol. 2, pp. 77–81, 2011.

150. J. Wei, L. Liao, Y. Xiao, P. Zhang, and Y. Shi, "Capture of carbon dioxide by amine-impregnated as-synthesized MCM-41," Journal of Environmental Sciences, vol. 22, no. 10, pp. 1558–1563, 2010.

151. Q. Wang, H. H. Tay, Z. Zhong, J. Luo, and A. Borgna, "Synthesis of high-temperature CO_2 adsorbents from organo-layered double hydroxides with markedly improved CO_2 capture capacity," Energy & Environmental Science, vol. 5, pp. 7526–7530, 2012.

152. H. Lin and B. D. Freeman, "Gas solubility, diffusivity and permeability in poly(ethylene oxide)," Journal of Membrane Science, vol. 239, no. 1, pp. 105–117, 2004. · ·

153. P. Chowdhury, C. Bikkina, and S. Gumma, "Gas adsorption properties of the chromium-based metal organic framework MIL-101," The Journal of Physical Chemistry C, vol. 113, no. 16, pp. 6616–6621, 2009.

154. P. Li, B. Ge, S. Zhang, S. Chen, Q. Zhang, and Y. Zhao, "CO_2 capture by polyethylenimine-modified fibrous adsorbent," Langmuir, vol. 24, no. 13, pp. 6567–6574, 2008.

155. B. Aziz, N. Hedin, and Z. Bacsik, "Quantification of chemisorption and physisorption of carbon dioxide on porous silica modified by propylamines: effect of amine density," Microporous and Mesoporous Materials, vol. 159, pp. 42–49, 2012.

156. M. M. Maroto-Valer, Z. Tang, and Y. Zhang, "CO_2 capture by activated and impregnated anthracites,"Fuel Processing Technology, vol. 86, no. 14-15, pp. 1487–1502, 2005.

157. X. Xu, C. Song, B. G. Miller, and A. W. Scaroni, "Adsorption separation of carbon dioxide from flue gas of natural gas-fired boiler by a novel nanoporous "molecular basket" adsorbent," Fuel Processing Technology, vol. 86, no. 14-15, pp. 1457–1472, 2005.

158. X. Xu, C. Song, J. M. Andresen, B. G. Miller, and A. W. Scaroni, "Novel polyethylenimine-modified mesoporous molecular sieve of MCM-41 type as high-capacity adsorbent for CO_2 capture," Energy and Fuels, vol. 16, no. 6, pp. 1463–1469, 2002.

159. J. Zhang, R. Singh, and P. A. Webley, "Alkali and alkaline-earth cation exchanged chabazite zeolites for adsorption based CO_2 capture," Microporous and Mesoporous Materials, vol. 111, no. 1–3, pp. 478–487, 2008.

160. H. R. Abid, H. Tian, H.-M. Ang, M. O. Tade, C. E. Buckley, and S. Wang, "Nanosize Zr-metal organic framework (UiO-66) for hydrogen and carbon dioxide storage," Chemical Engineering Journal, vol. 187, pp. 415–420, 2012.

161. C. Chen, J. Kim, and W.-S. Ahn, "Efficient carbon dioxide capture over a nitrogen-rich carbon having a hierarchical micro-mesopore structure," Fuel, vol. 95, pp. 360–364, 2012.

162. M. G. Plaza, C. Pevida, B. Arias et al., "Application of thermogravimetric analysis to the evaluation of aminated solid sorbents for CO_2 capture," Journal of Thermal Analysis and Calorimetry, vol. 92, no. 2, pp. 601–606, 2008.

163. A.-Y. Park, H. Kwon, A. J. Woo, and S.-J. Kim, "Layered double hydroxide surface modified with (3-aminopropyl) triethoxysilane by covalent bonding," Advanced Materials, vol. 17, no. 1, pp. 106–109, 2005.

164. N. S. Nhlapo, "Intercalation of fatty acids into layered double hydroxides," Tech. Rep., Department of Chemistery Faculty of Natural and Agricultural sciences South Africa, University of Pretoria, 2008.

165. M. S. Shafeeyan, W. M. A. Wan Daud, A. Houshmand, and A. Arami-Niya, "The application of response surface methodology to optimize the amination of activated carbon for the preparation of carbon dioxide adsorbents," Fuel, vol. 94, pp. 465–472, 2012.

166. M. Clausse, J. Merel, and F. Meunier, "Numerical parametric study on CO_2 capture by indirect thermal swing adsorption," International Journal of Greenhouse Gas Control, vol. 5, no. 5, pp. 1206–1213, 2011.

167. L. Wang, Z. Liu, P. Li, J. Yu, and A. E. Rodrigues, "Experimental and modeling investigation on post-combustion carbon dioxide capture using zeolite 13X-APG by hybrid VTSA process," Chemical Engineering Journal, vol. 197, pp. 151–161, 2012.

168. A. R. Kulkarni and D. S. Sholl, "Analysis of Equilibrium-Based TSA Processes for Direct Capture of CO_2 from Air," Industrial & Engineering Chemistry Research, vol. 51, pp. 8631–8645, 2012.

169. J. Merel, M. Clausse, and F. Meunier, "Experimental investigation on CO_2 post-combustion capture by indirect

187. D. Aaron and C. Tsouris, "Separation of CO_2 from flue gas: a review," Separation Science and Technology, vol. 40, no. 1–3, pp. 321–348, 2005.

188. A. Xu, A. Yang, S. Young, D. deMontigny, and P. Tontiwachwuthikul, "Effect of internal coagulant on effectiveness of polyvinylidene fluoride membrane for carbon dioxide separation and absorption,"Journal of Membrane Science, vol. 311, no. 1-2, pp. 153–158, 2008.

189. T.-L. Chew, A. L. Ahmad, and S. Bhatia, "Ordered mesoporous silica (OMS) as an adsorbent and membrane for separation of carbon dioxide (CO_2)," Advances in Colloid and Interface Science, vol. 153, no. 1-2, pp. 43–57, 2010.

190. C. A. Scholes, G. Q. Chen, G. W. Stevens, and S. E. Kentish, "Nitric oxide and carbon monoxide permeation through glassy polymeric membranes for carbon dioxide separation," Chemical Engineering Research and Design, vol. 89, no. 9, pp. 1730–1736, 2011. · ·

191. C. A. Scholes, S. E. Kentish, and G. W. Stevens, "Carbon dioxide separation through polymeric membrane systems for flue gas applications," Recent Patents on Chemical Engineering, vol. 1, pp. 52–66, 2008.

192. L. A. El-Azzami and E. A. Grulke, "Carbon dioxide separation from hydrogen and nitrogen by fixed facilitated transport in swollen chitosan membranes," Journal of Membrane Science, vol. 323, no. 2, pp. 225–234, 2008.

193. C. E. Powell and G. G. Qiao, "Polymeric CO_2/N_2 gas separation membranes for the capture of carbon dioxide from power plant flue gases," Journal of Membrane Science, vol. 279, no. 1-2, pp. 1–49, 2006.

194. L. A. El-Azzami and E. A. Grulke, "Carbon dioxide separation from hydrogen and nitrogen: facilitated transport in arginine salt-chitosan membranes," Journal of Membrane Science, vol. 328, no. 1-2, pp. 15–22, 2009.

195. G. Xomeritakis, C.-Y. Tsai, and C. J. Brinker, "Microporous sol-gel derived aminosilicate membrane for enhanced carbon

dioxide separation," Separation and Purification Technology, vol. 42, no. 3, pp. 249–257, 2005.

196. E. Favre, "Carbon dioxide recovery from post-combustion processes: can gas permeation membranes compete with absorption?" Journal of Membrane Science, vol. 294, no. 1-2, pp. 50–59, 2007.

197. A. Julbe, "Chapter 6 Zeolite membranes—synthesis, characterization and application," Studies in Surface Science and Catalysis, vol. 168, pp. 181–219, 2007.

198. D. W. Shin, S. H. Hyun, C. H. Cho, and M. H. Han, "Synthesis and CO_2/N_2 gas permeation characteristics of ZSM-5 zeolite membranes," Microporous and Mesoporous Materials, vol. 85, no. 3, pp. 313–323, 2005.

199. M. Anderson and Y. S. Lin, "Carbonate-ceramic dual-phase membrane for carbon dioxide separation,"Journal of Membrane Science, vol. 357, no. 1-2, pp. 122–129, 2010.

200. D. Shekhawat, D. R. Luebke, and H. W. Pennline, "A review of carbon dioxide selective membranes," A Topical Report DOE/NETL-2003/1200, Department of Energy, National Energy Technology Laboratory, 2003.

201. P. Kumar, S. Kim, J. Ida, and V. V. Guliants, "Polyethyleneimine-modified MCM-48 membranes: effect of water vapor and feed concentration on N_2/CO_2 selectivity," Industrial and Engineering Chemistry Research, vol. 47, no. 1, pp. 201–208, 2008.

202. T. C. Merkel, H. Lin, X. Wei, and R. Baker, "Power plant post-combustion carbon dioxide capture: an opportunity for membranes," Journal of Membrane Science, vol. 359, no. 1-2, pp. 126–139, 2010.

203. Y. Cai, Z. Wang, C. Yi, Y. Bai, J. Wang, and S. Wang, "Gas transport property of polyallylamine-poly(vinyl alcohol)/polysulfone composite membranes," Journal of Membrane Science, vol. 310, no. 1-2, pp. 184–196, 2008.

222. A. L. B. Ahmad, Z. A. Jawad, S. C. Low, and H. S. Zein, "Prospect of mixed matrix membrane towards CO_2 Separation," Journal of Membrane Science & Technology, vol. 2, article e110, 2012.

223. C. A. Scholes, G. Q. Chen, G. W. Stevens, and S. E. Kentish, "Plasticization of ultra-thin polysulfone membranes by carbon dioxide," Journal of Membrane Science, vol. 346, no. 1, pp. 208–214, 2010.

224. N. Du, H. B. Park, G. P. Robertson et al., "Polymer nanosieve membranes for CO_2-capture applications," Nature Materials, vol. 10, no. 5, pp. 372–375, 2011.

225. P. Uchytil, J. Schauer, R. Petrychkovych, K. Setnickova, and S. Y. Suen, "Ionic liquid membranes for carbon dioxide-methane separation," Journal of Membrane Science, vol. 383, no. 1-2, pp. 262–271, 2011.

226. O. G. Nik, X. Y. Chen, and S. Kaliaguine, "Amine-functionalized zeolite FAU/EMT-polyimide mixed matrix membranes for CO_2/CH_4 separation," Journal of Membrane Science, vol. 379, no. 1-2, pp. 468–478, 2011.

227. Y. C. Hudiono, T. K. Carlisle, J. E. Bara, Y. Zhang, D. L. Gin, and R. D. Noble, "A three-component mixed-matrix membrane with enhanced CO_2 separation properties based on zeolites and ionic liquid materials," Journal of Membrane Science, vol. 350, no. 1-2, pp. 117–123, 2010.

228. A. Kovvali and G. Obuskovic, "Immobilized liquid membranes for CO_2 separation'," in Proceedings of the Preprints of Symposia-American Chemical Society, pp. 665–667, Division of Fuel Chemistry, American Chemical Society, 2000.

229. Z. Wang, L. E. K. Achenie, S. J. Khativ, and S. T. Oyama, "Simulation study of single-gas permeation of carbon dioxide and methane in hybrid inorganic-organic membrane," Journal of Membrane Science, vol. 387-388, no. 1, pp. 30–39, 2012.

230. S.-P. Yan, M.-X. Fang, W.-F. Zhang et al., "Experimental study on the separation of CO_2 from flue gas using hollow fiber

membrane contactors without wetting," Fuel Processing Technology, vol. 88, no. 5, pp. 501–511, 2007.

231. J.-L. Li and B.-H. Chen, "Review of CO_2 absorption using chemical solvents in hollow fiber membrane contactors," Separation and Purification Technology, vol. 41, no. 2, pp. 109–122, 2005.

232. Y.-S. Kim and S.-M. Yang, "Absorption of carbon dioxide through hollow fiber membranes using various aqueous absorbents," Separation and Purification Technology, vol. 21, no. 1-2, pp. 101–109, 2000.

233. K. Sugiura, K. Takei, K. Tanimoto, and Y. Miyazaki, "The carbon dioxide concentrator by using MCFC," Journal of Power Sources, vol. 118, no. 1-2, pp. 218–227, 2003.

234. H. Herzog, "Assessing the feasibility of capturing CO_2 from the air," Tech. Rep., MIT Laboratory for Energy and the Environment, Massachusetts Institute of Techology, Cambridge, Mass, USA, 2003.

235. M. R. M. Abu-Zahra, J. P. M. Niederer, P. H. M. Feron, and G. F. Versteeg, "CO_2 capture from power plants. Part II. A parametric study of the economical performance based on mono-ethanolamine,"International Journal of Greenhouse Gas Control, vol. 1, no. 2, pp. 135–142, 2007.

2

Chemical Looping Combustion of Methane: A Technology Development View

Rutuja Bhoje[1], Ganesh R. Kale[1], Nitin Labhsetwar[2], and Sonali Borkhade[1]

[1]CSIR-National Chemical Laboratory (CSIR-NCL), Pune 411008, India

[2]CSIR-National Environmental Engineering Research Institute (CSIR-NEERI), Nagpur 440020, India

ABSTRACT

Methane is a reliable and an abundantly available energy source occurring in nature as natural gas, biogas, landfill gas, and so forth.

Clean energy generation using methane can be accomplished by using chemical looping combustion. This theoretical study for chemical looping combustion of methane was done to consider some key technology development points to help the process engineer choose the right oxygen carrier and process conditions. Combined maximum product $(H_2O + CO_2)$ generation, weight of the oxygen carrier, net enthalpy of CLC process, byproduct formation, CO_2 emission from the air reactor, and net energy obtainable per unit weight (gram) of oxygen carrier in chemical looping combustion can be important parameters for CLC operation. Carbon formed in the fuel reactor was oxidised in the air reactor and that increased the net energy obtainable from the CLC process but resulted in CO_2 emission from the air reactor. Use of $CaSO_4$ as oxygen carrier generated maximum energy (-5.3657 kJ, 800°C) per gram of oxygen carrier used in the CLC process and was found to be the best oxygen carrier for methane CLC. Such a model study can be useful to identify the potential oxygen carriers for different fuel CLC systems.

INTRODUCTION

Energy demand and thus energy generation are ever increasing in many parts of the world. Carbon-based fossil fuels are the main source of energy for combustion reactions. However, the major product of energy generation using fossil fuels by combustion is CO_2, and growing CO_2 pollution has become a matter of serious concern amongst developed as well as developing countries. CO_2 is mainly emitted from energy sector that uses coal, oil, and natural gas for combustion to generate energy for different purposes using air as an oxygen source. Although CO_2 capture and sequestration is being projected as a potential option to control GHG emissions, CO_2 capture technologies in-vogue are beset with several limitations including cost and energy penalty [1, 2]. Hence, these flue gases are directly vented to the atmosphere without CO_2 separation and became responsible for environmental impacts of energy generation from fossil fuels, for example, global warming and climate change phenomena [3–5]. In 2010, CO_2 emissions have already increased to 389.0 ppm and the burning of fossil fuels is one of the main causes for the same as reported by World Meteorological Organization [6]. This has promoted the development of clean energy technologies as a major research area worldwide. Chemical looping combustion (CLC) is an emerging clean energy technology to generate energy using abundantly available fuels and also provides sequestration ready CO_2 stream [7, 8]. CLC uses a solid oxygen carrier (OC) to oxidize the carbon and hydrogen in the fossil fuel to CO_2 and H_2O in an endothermic fuel reactor. The reduced OC is regenerated by air oxidation in the exothermic air reactor [9, 10] in the next step. Both reactors are interconnected and operate simultaneously. The energy released in the CLC system is considered to be of similar magnitude as direct combustion but with a crucial advantage of having separate CO_2 stream [11, 12]. The pure CO_2 steam can be captured/sequestered easily, thereby reducing CO_2 emissions to atmosphere [13].

Development of chemical looping combustion technology requires the relatively cheap and long term availability of fuel and oxygen carrier. Fossil fuel such as methane is abundantly available as natural gas in many countries [14–21]. It is also available as biogas [22–25], landfill gas [26–31], gas hydrates [32–35], and coal bed methane [36] and can be generated from other organic wastes [37–40], and it is a major source of energy by combustion worldwide [41]. The selection of a suitable oxygen carrier (OC) for methane CLC is one the most important aspects for successful technology development study. Apart from the low-cost and abundant availability criteria, the OC must have suitable reactivity with the fuel, easy reducibility, and regenerability under CLC conditions in addition to some other essential properties such as high melting point, attrition resistance, mechanical stability, negligible loses during operation, negligible toxicity, and environmental impacts. The suitability of the oxygen carrier for a fuel can be determined by trial experiments and also by theoretical studies. Some theoretical (thermodynamic) studies of chemical looping combustion have already been reported [42–44], while experimental research

mixed trend; that is, for some oxygen carriers, the H_2O formation increased with increase in temperature, while in other cases, the H_2O yield decreased with increase in temperature. From the figure, it was seen that the moles of H_2O slightly decreased from 1.85 to 1.83 moles with increase in temperature from 600°C to 1200°C for NiO. The same trend was observed for CoO; that is, it decreased from 1.705 to 1.62 moles. But the H_2O yield increased from 1.80 to 1.89 moles up to 850°C then it decreased to 1.81 moles (1200°C) for $CaSO_4$, while the H_2O yield increased with increase in temperature (i.e., 1.16 at 600°C to 1.64 at 1200°C) for Na_2SO_4. It was also observed that the H_2O yield first decreased from 0.83 to 0.78 moles up to 750°C and then increased to 1.03 moles for Fe_2O_3. However, the H_2O yield remained almost constant at all temperatures (that was the maximum moles of H_2O obtainable) for CuO when compared to all other oxygen carriers (e.g., 1.99 moles at 600°C and 1.99 moles at 1200°C). But the H_2O yield decreased from 0.87 to 0.29 moles with increase in temperature from 600 to 1200°C for Mn_2O_3. The minimum moles of H_2O produced were 0.29 moles for Mn_2O_3 at 1200°C, while the maximum H_2O moles produced were 2.00 for CuO at 600°C. The ranking of oxygen carriers with respect to increasing order of moles of H_2O produced from methane was found to be as follows: CuO (1.99, 600°C) > $CaSO_4$ (1.89, 850°C) > NiO (1.86, 600°C) > CoO (1.70, 600°C) > Na_2SO_4 (1.64, 1200°C) > Fe_2O_3 (1.03, 1200°C) > Mn_2O_3 (0.87, 600°C).

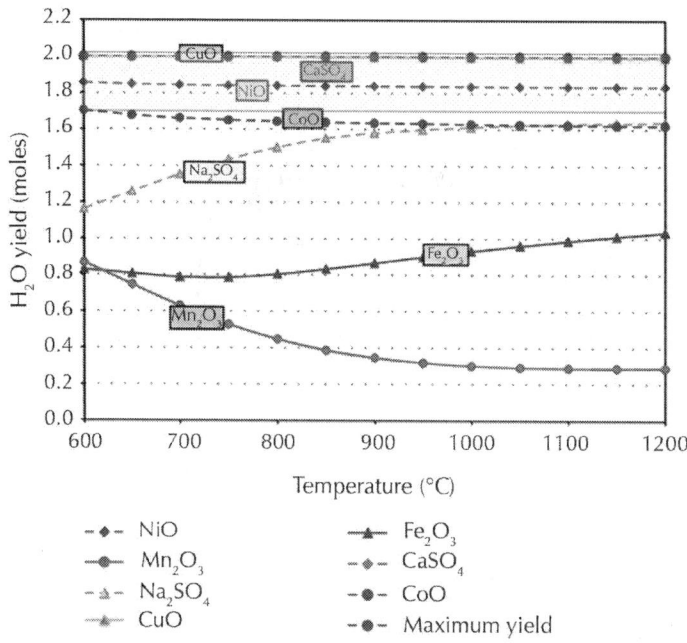

Figure 1: H_2O yield for CLC of methane.

CO₂ Yield

Similarly, the effect of temperature and oxygen carrier on CO_2 formation in the CLC fuel reactor was studied. The net CO_2 obtainable in the CLC fuel reactor was 1 mole per mole of methane feed. The trends in the CO_2 formation with selected oxygen carriers are shown in Figure 2. It was observed that the CO_2 yield also showed mixed trends (similar to H_2O) for different oxygen carriers with increase in temperature from 600 to 1200°C. The maximum CO_2 yield in CLC fuel reactor was 1.00 mole for its reaction with CuO at 600°C, while the minimum CO_2 yield of 0.06 moles for Mn_2O_3 was calculated at 1200°C. The selectivity of the oxygen carriers based on maximum CO_2 yield (moles) was analysed and was found to be as follows: CuO (1.00, 600°C) > NiO (0.96, 600°C) > $CaSO_4$ (0.95, 900°C) > CoO (0.88, 600°C) > Na_2SO_4 (0.71, 1050°C) > Fe_2O_3 (0.38, 800°C) > Mn_2O_3 (0.21, 700°C).

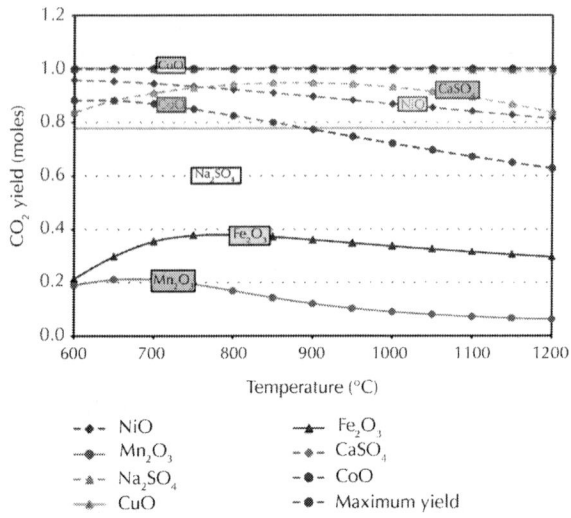

Figure 2: CO_2 yield for CLC of methane.

(H₂O + CO₂) Yield

It was clearly seen that the ranking of the oxygen carriers with their suitable operating temperatures was not the same for both major products. Hence, selection of the optimum process conditions was done by using a different parameter. In this case, the combined desired product (H_2O and CO_2) yield from the CLC fuel reactor is considered as a better parameter than individual yields. The maximum yield of 3 moles (1 mole CO_2 + 2 moles H_2O) can be obtained from one mole methane in the CLC fuel reactor. The combined yields of H_2O and CO_2 were plotted in Figure 3 for selected oxygen carriers in the considered temperature range. It was observed that similar to the individual H_2O and CO_2 yields, the oxygen carriers showed different trends for combined yield ($CO_2 + H_2O$) with increase in CLC fuel reactor temperature. It was observed that the ($CO_2 + H_2O$) yield for NiO decreased from 2.81 to 2.65 moles with increase in temperature from 600 to 1200°C as reported earlier [42]. Similar trend was observed for Mn_2O_3 and CoO, where the ($CO_2 + H_2O$) yield decreased from 2.59 to 2.25

moles for CoO and from 1.06 moles to 0.35 moles for Mn_2O_3 with increase in temperature from 600 to 1200°C. The $(CO_2 + H_2O)$ yield increased from 1.45 to 2.34 moles up to 1100°C, and then it remained almost constant till 1200°C for Na_2SO_4. The $(CO_2 + H_2O)$ yield increased with increase in temperature, that is, from 1.04 moles (600°C) to 1.33 moles (1200°C) for Fe_2O_3, while for $CaSO_4$, the $(CO_2 + H_2O)$ yield increased from 2.64 to 2.84 moles up to 850°C and then decreased to 2.65 moles at 1200°C. In case of CuO, the $(CO_2 + H_2O)$ yield remained almost constant at all temperatures, that is, ~3.00 moles. A maximum $(CO_2 + H_2O)$ yield of 3.00 moles was obtained with CuO at almost all temperatures, while the minimum $(CO_2 + H_2O)$ yield of 0.35 moles was observed for Mn_2O_3 at 1200°C. The ranking of oxygen carriers based on the maximum $(CO_2 + H_2O)$ yield was seen as follows: CuO (2.99, 600°C) > $CaSO_4$ (2.84, 850°C) > NiO (2.81, 600°C) > CoO (2.59, 600°C) > Na_2SO_4 (2.34, 1100°C) > Fe_2O_3 (1.33, 1200°C) > Mn_2O_3(1.06, 600°C). This combined product criterion for selection of oxygen carriers and process temperature was useful for process operation.

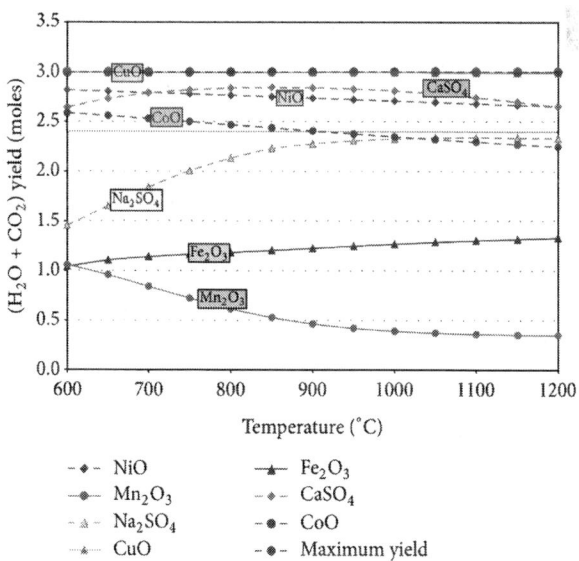

Figure 3: $(H_2O + CO_2)$ yield for CLC of methane.

Undesired Product Formation

Some undesired products are also formed in the CLC fuel reactor due to side reactions and thermodynamic limitations under the CLC conditions. An estimation of the nature and amount of the byproducts is important for technology development and hence was studied in detail.

Syngas (H_2 + CO) Formation

Syngas (H_2 + CO) is one of the undesired products of the CLC process. The data of syngas formation in the CLC fuel reactor with selected oxygen carriers within the temperature range from 600 to 1200°C was analyzed and is shown in Figure 4. From the figure, it was observed that the syngas formation increased with increase in process temperature from 600 to 1200°C for all oxygen carriers except for Fe_2O_3, where it was slightly decreased at higher temperature. The syngas formation (moles) increased from 1.13 to 2.64 (Mn_2O_3), 0.35 to 0.75 (CoO), 0.17 to 0.35 (NiO), 0.16 to 0.49 (Na_2SO_4), 0.05 to 0.29 ($CaSO_4$), and 0.00 to 0.01 moles (CuO) with increase in temperature from 600 to 1200°C. But it was also seen that the syngas formation increased from 1.16 to 1.75 moles up to 900°C and then slightly decreased to 1.67 moles at 1200°C in the case of Fe_2O_3. CuO produced negligible syngas when compared to all other oxygen carriers at all temperatures. The minimum syngas moles produced were negligible for CuO, while the maximum syngas of 2.64 moles was produced by Mn_2O_3 at 1200°C in this study. The ranking of the oxygen carriers based on minimum syngas production (moles) in the CLC fuel reactor was observed as follows: CuO (0.00) > $CaSO_4$ (0.05) > Na_2SO_4 (0.16) > NiO (0.17) > CoO (0.35) > Mn_2O_3 (1.13) > Fe_2O_3 (1.16) at 600°C.

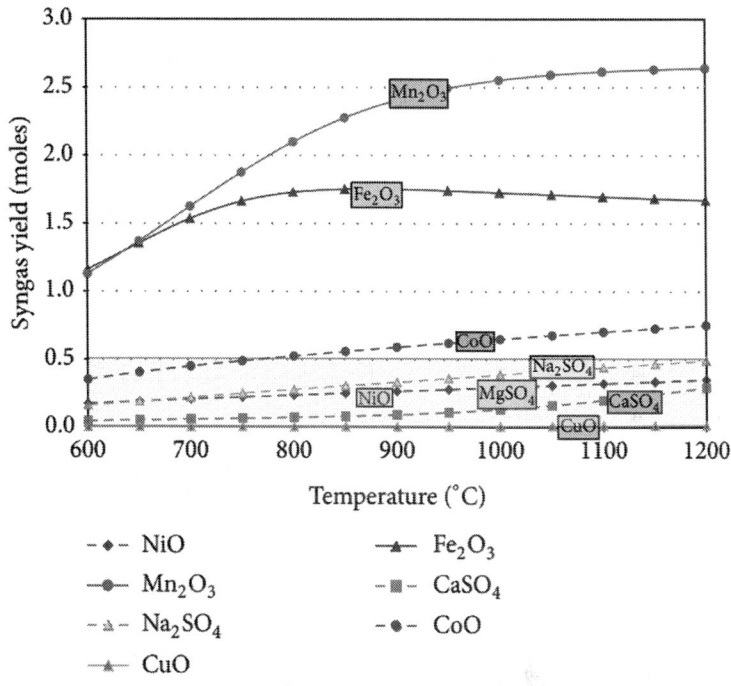

Figure 4: Syngas formation for CLC of methane.

Carbon Formation

Carbon formation is undesirable in the CLC process as coking may reduce the activity of the solid oxygen carriers. Figure 5 shows the carbon formation in the CLC fuel reactor for methane fuel within the temperature range of 600°C to 1200°C at 1 bar pressure. From the figure, it was observed that the carbon formation was higher at lower temperatures (till ~700°C) due to carbon formation side reactions (5), (6), and (7). It was also seen that the carbon formation decreased with increase in temperature from 600 to 1200°C for all oxygen carriers except for CuO (as found earlier [1]) and $CaSO_4$ as it was almost zero at all temperatures. The carbon formation steeply decreased from 0.68 at 600°C to 0.010 at 1200°C for Mn_2O_3, while it decreased from 0.63 to 0.001 moles for Fe_2O_3. The carbon formation slightly decreased with increase in temperature, that is, from 0.06 moles at 600°C to 0.00 at 1200°C for CoO. The carbon formation also decreased from 0.013 moles to 0.00 (NiO), from 0.003 moles to 0.00 (Na_2SO_4) within 600 to 1200°C. Fe_2O_3 and Mn_2O_3 produced higher carbon in comparison with the rest of the oxygen carriers at the same temperatures. The maximum carbon yield was 0.68 moles with Mn_2O_3 at 600°C. This solid carbon can be separated from gas stream but cannot be separated from the oxygen carrier, and it is carried to the air reactor where it is oxidised to CO_2. This combustion generates energy but also contaminates the N_2 stream from the air reactor with some CO_2 and increases the air requirement in CLC air reactor. It was seen that all oxygen carriers gave their maximum carbon yield at 600°C. The selectivity of the oxygen carrier as per the decreasing yield of carbon (in moles) is observed as follows: Mn_2O_3 (0.677) < Fe_2O_3 (0.625) < CoO (0.06) < NiO (0.013) < Na_2SO_4 (0.003) < $CaSO_4$ (0.0002) < CuO (0.00) at 600°C.

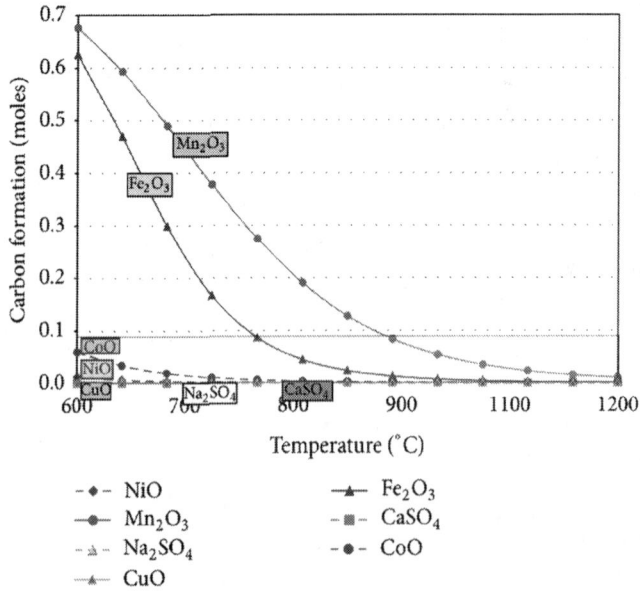

Figure 5: Carbon formation for CLC of methane.

SO_2 and H_2S Formation

The possibility of SO_2, COS, and H_2S formation in the CLC fuel reactor while using sulphates as oxygen carriers was also considered in this study, as it was reported by other researchers [79, 80]. Both SO_2 and H_2S are highly undesirable for the process. The results showed that SO_2 and H_2S formation takes place in the CLC fuel reactor.

Figure 6 shows the H_2S and SO_2 formation for the 3 sulphate-based oxygen carriers within the temperature range from 600 to 1200°C at 1 bar pressure per mole methane used in the feed. It was observed that the H_2S formation decreased with increase in temperature for all sulphates. The H_2S formation (moles) decreased from 0.16 to 0.06 for $CaSO_4$ and from 0.69 to 0.08 moles for Na_2SO_4 with increase in temperature from 600 to 1200°C. The maximum H_2S yield was found to be 0.69 moles at 600°C for Na_2SO_4, while the minimum H_2S yield was found to be 0.06 moles at 1200°C for $CaSO_4$. Figure 6 also showed that the SO_2 formation increased with increase in temperature for all sulphates. The SO_2 formation increased from 0.00 to 0.1 moles for $CaSO_4$ and from 0.00 to 0.02 moles for Na_2SO_4 with increase in temperature from 600 to 1200°C. The maximum SO_2 yield was found to be 0.1 moles at 1200°C for $CaSO_4$, while zero SO_2 formation was observed at lower temperatures till 850°C for Na_2SO_4 and till 650°C for $CaSO_4$. Hence, according to the data analysis, the choice of better oxygen carrier amongst sulphates was $CaSO_4 > Na_2SO_4$. H_2S or SO_2 formation reduces the sulphate to its oxide (e.g., CaO and Na_2O), which can be regenerated by using an acid but may require additional processing cost. Commercial systems to trap SO_2 as well as H_2S are available; however, deterioration of sulphate-based oxygen carrier is always a matter of greater concern. Depending on the cost and availability of the sulphate, smart choice of sulphate oxygen carrier can be made.

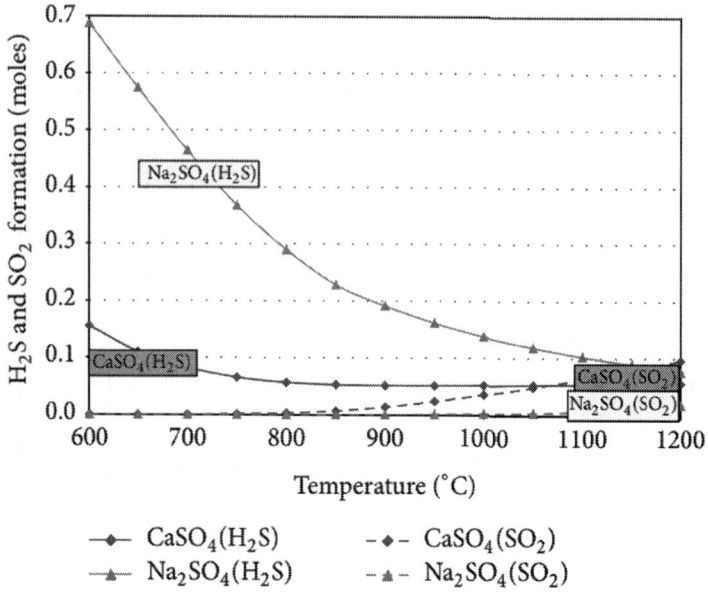

Figure 6: H$_2$S and SO$_2$ gases formation for CLC of methane.

Reactant Conversions

A study of conversion of the reactants is also important for CLC technology development and hence studied in detail and discussed in the following section.

CH$_4$ Conversion

Methane conversion in the CLC fuel reactor is one of the important aspects for choosing the oxygen carrier and operating temperature range, as the reactivity of oxygen carriers with methane may be different. Figure 7depicts the trends in CH$_4$ conversion with increase in temperature from 600 to 1200°C at 1 bar pressure for the selected oxygen carriers. It was observed from Figure 7 that the CH$_4$ conversion was almost 100% for all oxygen carriers except for Fe$_2$O$_3$ and Mn$_2$O$_3$, where lower methane conversion was seen at lower temperatures. The CH$_4$ conversion increased from 94.22%

to 100% for Fe_2O_3, while it was increased from 95.46% to 100% for Mn_2O_3 (also experimentally demonstrated earlier [81]), with increase in CLC fuel reactor temperature from 600 to 1200°C. Complete conversion of CH_4 was achieved for all oxygen carriers in the conditions considered in this study, as these oxygen carriers gave almost 100% methane conversion above 800°C only.

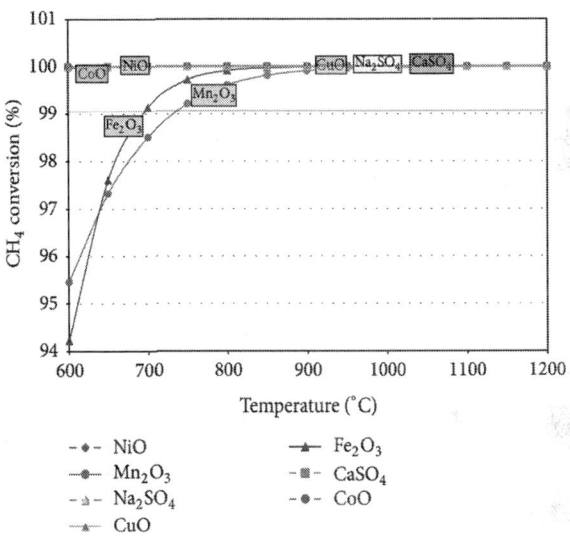

Figure 7: CH_4 conversion for CLC of methane.

Oxygen Carrier Reduction

The reduction of oxygen carrier in the fuel reactor is a key reaction responsible for oxygen availability in CLC reaction and is controlled by the thermodynamic limitations. A study of oxygen carrier conversions in the fuel reactor is also important for calculating the air requirement for regeneration of the oxygen carrier. Hence, all possible reduced states of the oxygen carrier were considered in this study. It was observed that oxygen carriers like NiO, CuO, and CoO were reduced to their respective metals in the fuel reactor. However, oxygen carriers like Fe_2O_3 and Mn_2O_3 were reduced to different reduced states. Similarly, many other species were also formed with the sulfate oxygen carriers in the CLC fuel reactor. The species formed in the CLC fuel reactor of Fe_2O_3 and $CaSO_4$ were discussed in detail.

Reduced Species of Fe_2O_3

Figure 8(a) shows the different reduced species of Fe_2O_3 for CLC of methane in the temperature range from 600 to 1200°C at 1 bar pressure. From the figure, it was observed that Fe_2O_3 was almost reduced completely into its different reduced species like Fe, FeO, and Fe_3O_4. The moles of FeO and Fe produced increased with increase in temperature; that is, for FeO, it increased from 0.99 to 1.31 moles and for Fe, it increased from 0.44 to 1.08 moles, while the moles of Fe_3O_4 produced decreased with increase in temperature; that is, it decreased from 0.41 to 0.09 moles, while negligible amount of Fe_2O_3 was observed remaining at all temperatures.

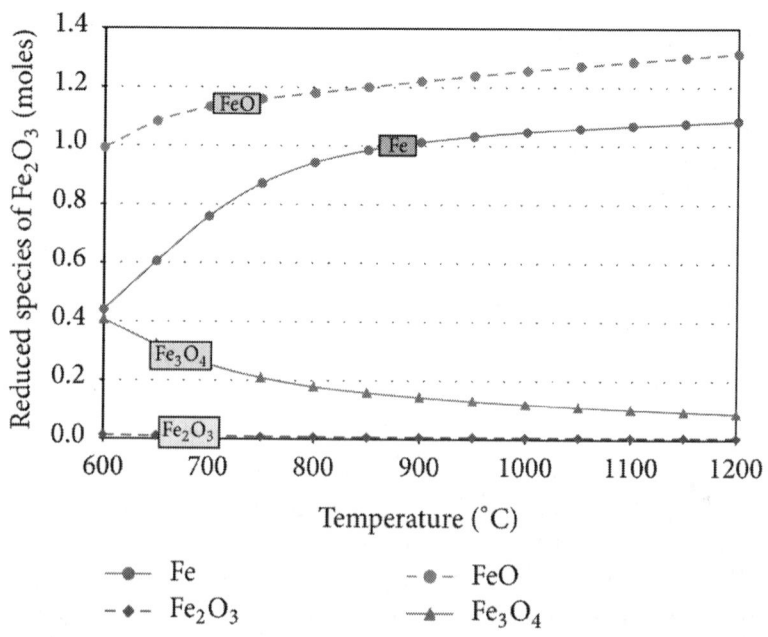

(a)

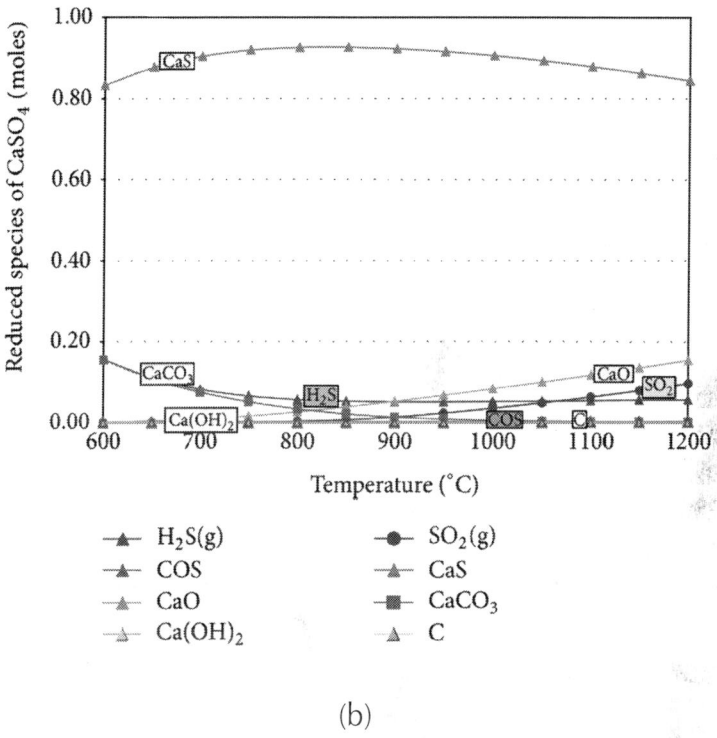

(b)

Figure 8: (a) Reduced species of Fe_2O_3 for CLC of methane. (b) Reduced species of $CaSO_4$ for CLC of methane.

Reduced Species of CaSO$_4$

Figure 8(b) shows the different reduced species of $CaSO_4$ for CLC of methane process in the temperature range from 600 to 1200°C at 1 bar pressure. Reduction behaviour of sulphur containing oxygen carrier is even more important due to the possibility of formation of sulphur-based compounds. From Figure 8(b), it was observed that $CaSO_4$ almost reduced completely into its different species like CaS, CaO, $CaCO_3$, $Ca(OH)_2$, COS, H_2S, and SO_2. The moles of SO_2 and CaO produced were increased with increase in temperature; that is, it increased from 0.00 to 0.1 moles for SO_2 and increased from 0.00 to 0.15 moles for CaO, while moles of $CaCO_3$ produced decreased with increase in temperature; that is, it decreased from

0.15 moles to its lowest value. Moles of H_2S produced decreased till 950°C, and then slightly increased with increase in temperature; that is, it decreased from 0.16 to 0.05 moles and then increased to 0.06 moles while reverse trend was observed for CaS, that is, moles of CaS produced increased till 850°C and then slightly decreased with increase in temperature, for example, it increased from 0.83 to 0.93 moles and then decreased to 0.84 moles. CaS being the most desirable reduced state, the CLC fuel reactor temperature can be limited to 950°C when $CaSO_4$ is used as oxygen carrier. Negligible amount of COS and $Ca(OH)_2$ formation was observed with increase in temperature.

It was observed that the sulfate oxygen carriers suffer losses due to conversion to the oxide and carbonate in the CLC fuel reactor. The oxide and carbonate cannot be separated from the solid stream in continuous operation. Hence, a small amount of sulfate oxygen carrier needs to be added to the process continuously to keep the amount of active oxygen carrier intact. This will also require a continuous small purge of the spent oxygen carrier from the system to avoid overloading of the system. This purged oxygen carrier can be reacted with dilute H_2SO_4, filtered, and dried to convert the oxide and carbonate back to the sulfate for reuse in the system. This modification will be governed by the throughput of the CLC plant, operating cost, and cost of the oxygen carrier.

Process Energy

The CLC system consists of two reactors in which energy and oxygen carrier transfer occurs between the exothermic (air) reactor and the endothermic (fuel) reactor. It is generally considered that the magnitude of energy obtained in CLC is same as the energy of combustion of the fuel. However, the energy obtainable in CLC depends on the conversion of fuel to different products at different temperatures. As seen in the earlier sections, some other byproducts are also formed in the CLC fuel reactor apart from CO_2 and H_2O. This affects the net energy obtainable from the CLC system for the different oxygen carriers. The study of fuel and air reactor enthalpy trends within the selected temperature range from 600 to 1200°C was essential to calculate the optimum conditions and is discussed in the next section.

Enthalpy of the Fuel Reactor

Methane undergoes oxidation in the fuel reactor generating primarily CO_2 and H_2O, and the oxygen carrier is reduced to its lower oxidation state. The enthalpy of this reaction is plotted against temperature in Figure 9 for different oxygen carriers, based on the equilibrium compositions obtained at those conditions. The CLC fuel reactor enthalpy showed different trends for different oxygen carriers. It was observed that the fuel reactor enthalpy increased with increase in temperature from 600 to 1200°C for CoO (135.15 kJ to 156.22 kJ), while it increased till ~1100°C and then became almost constant with further increase in temperature for Fe_2O_3, Mn_2O_3, and $CaSO_4$; that is, it increased from 150.09 kJ to 263.04 kJ and remained almost constant (Fe_2O_3), increased from 30.66 kJ to 124.41 kJ and remained almost constant (Mn_2O_3), and also increased from 137.42 kJ to 161.80 kJ (from 600 to 1150°C) and remained almost constant for $CaSO_4$. However, the reaction enthalpy decreased with increase in temperature in the case of NiO, where it decreased from 146.33 kJ (600°C) to 139.61 kJ (1200°C). The reaction enthalpy showed some increase-decrease trend with increase in temperature for CuO and Na_2SO_4 as shown in the figure. The ranking of oxygen carriers based on the minimum fuel reactor enthalpy is as follows: Fe_2O_3 (263.04 kJ, 1100°C) < Na_2SO_4 (194.90 kJ, 1200°C) < $CaSO_4$ (161.80 kJ, 1150°C) < CoO (156.22 kJ, 1200°C) < NiO (146.33 kJ, 600°C) < Mn_2O_3 (124.41 kJ, 1100°C) < CuO (−161.48 kJ, 1100°C). However, it was observed that some of the oxygen carriers did not get completely converted to the desired reduced state, and hence there was some loss of oxygen carrier material. Table 3 shows the conversion of the oxygen carrier to its lower oxidation states, some of which can be regenerated by acid reaction. However, this loss was found to be negligible, and assuming that the oxygen carriers are very cheap, the CLC process did not seem to suffer any major cost issues on this point.

Table 3: Conversion of the oxygen carriers during methane CLC

Sr. number	Oxygen carrier	Initial moles	Reduced state	Waste
1	CuO	4	Cu (4 moles) 100%	
2	NiO	4	Ni (3.80 mole) 95.05%	
3	CoO	4	Co (3.51 mole) 87.75%	
4	Fe_2O_3	1.33	Fe (1.08 mole) 40.55%, FeO (1.31 mole) 49.14% Fe_3O_4(0.09 mole) 3.29%	
5	Mn_2O_3	1.33	MnO (2.68 mole) 99.99%	
6	$CaSO_4$	1	CaO (0.04 mole) 3.88% CaS (0.93 mole) 92.55% $CaCO_3$(0.02 mole) 2.15%	CaO (0.04 mole) 3.88% $CaCO_3$(0.02 mole) 2.15%
7	Na_2SO_4	1	Na_2S (0.87 mole) 86.677% Na_2CO_3(0.033 mole) 3.23% NaOH (0.02 mole 1.69%)	Na_2CO_3(0.033 mole) 3.23%

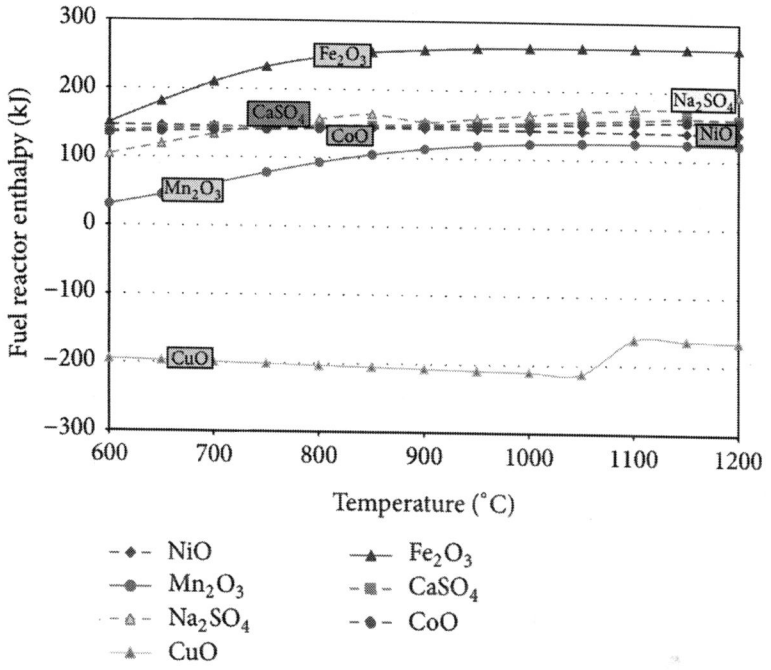

Figure 9: Fuel reactor enthalpy for CLC of methane.

Enthalpy of the Air Reactor

The reduced oxygen carrier along with solid carbon is continuously circulated back to the air reactor for regeneration. Air is passed through the air reactor, and the reduced oxygen carrier and carbon get oxidized completely in the air reactor. This carbon oxidation to CO_2 in the air reactor increases the exothermicity of the air reactor as shown below:

$$C + O_2 \longrightarrow CO_2 \quad \Delta H = -394.75 \, kJ \quad (800°C)$$

$$(10)$$

Figure 10 shows the variation of air reactor enthalpy within the temperature range from 600 to 1200°C at 1 bar pressure. It was observed that exothermicity of oxidation increased with increase in air reactor temperature for Na_2SO_4; that is, it increased from −268.51 to −787.29 kJ for Na_2SO_4, while the exothermicity decreased with increase in temperature for NiO, CoO, and Mn_2O_3; it decreased from −903.24 kJ to −852.05 kJ for NiO, decreased from −845.94 kJ to −762.33 kJ for CoO, and decreased from −513.18 kJ to −240.21 kJ for Mn_2O_3 with increase in air reactor temperature from 600 to 1200°C. A different trend was observed for $CaSO_4$, where the exothermicity of oxidation first increased and then decreased with increase in temperature; that is, it increased from −794.68 kJ (600°C) to −877.44 kJ (800°C) and then decreased to −784.08 kJ (1200°C), while mixed trend was observed for CuO. An enthalpy decrease-increase-decrease trend was observed for Fe_2O_3 as seen in the figure. The selectivity of oxygen carriers according to the maximum exothermicity of the reaction is as follows: NiO (−903.24 kJ, 600°C) > $CaSO_4$ (−877.44 kJ, 800°C) > CoO (−845.94 kJ, 600°C) > Na_2SO_4 (−787.29 kJ, 1200°C) > CuO (−640.17 kJ, 1100°C) > Fe_2O_3 (−629.23 kJ, 1200°C) > Mn_2O_3 (−513.18 kJ, 600°C).

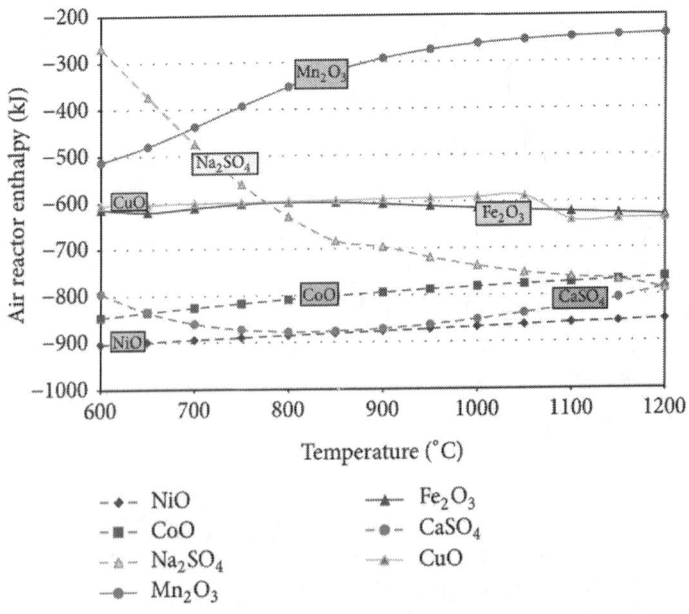

Figure 10: Air reactor enthalpy for CLC of methane.

Net Process Heat

The net energy required for the CLC process was calculated by summing up the energies of the exothermic air reactor and endothermic fuel reactor for identical oxygen carrier and carbon processing for 1 mole methane feed. The energy generated due to the exothermic air reactor is utilized for reduction of the oxygen carrier in the fuel reactor. The carbon formed in the fuel reactor and its oxidation to CO_2 in the air reactor provided additional energy.

Figure 11 shows that net process heat was obtained by using different oxygen carriers within the temperature range from 600 to 1200°C at 1 bar pressure. It was observed that the net process exothermicity decreased with increase in temperature for NiO, Mn_2O_3, and CoO; that is, it decreased from −756.91 kJ to −712.44 kJ for NiO, decreased from −482.52 kJ to −116.99 kJ for Mn_2O_3, and decreased from −710.79 kJ to −606.11 kJ for CoO, while net process heat increased with increase in temperature for Na_2SO_4; that is, it increased from −164.35 kJ to −592.38 kJ for Na_2SO_4. The net exothermicity first increased and then decreased with increase in temperature; that is, it increased from −657.26 kJ (600°C) to −730.48 kJ (800°C) and then decreased to −623.75 kJ (1200°C) for $CaSO_4$, while the exothermicity first decreased and then increased with increase in temperature for Fe_2O_3; that is, it decreased from −464.15 kJ (600°C) to −346.2 kJ (900°C) and then increased to −366.46 kJ (1200°C). It was observed that the net process exothermicity remained almost constant for CuO. The oxygen carriers were ranked according to the increase in the net process exothermicity (kJ), and the trend was observed as follows: CuO (−801.68 kJ, 1150°C) > NiO (−756.91 kJ, 600°C) > $CaSO_4$ (−730.48 kJ, 800°C) > CoO (−710.79 kJ, 600°C) > Na_2SO_4 (−592.38 kJ, 1200°C) > Mn_2O_3 (−482.52 kJ, 600°C) > Fe_2O_3 (−464.15 kJ, 600°C). These values were compared with the value of energy of methane combustion. It was observed that the CLC did not exactly give the complete energy that can be obtained from 1 mole methane by air oxidation. This is due to the byproduct formation in the fuel reactor. Although carbon formed in the CLC fuel reactor is oxidized in the air reactor, the syngas and unconverted methane escape in the CLC fuel reactor product stream mainly causing the loss in fuel and subsequent energy. This loss can be reduced by adding excess of oxygen carrier in the fuel reactor. However, this large additional amount of oxygen carrier for small conversion in the fuel reactor thereby increases the process operating cost. Hence, adding a secondary fuel reactor using some more oxygen carriers to oxidise the H_2, CO, and CH_4 completely in the second stage seems to be a better option.

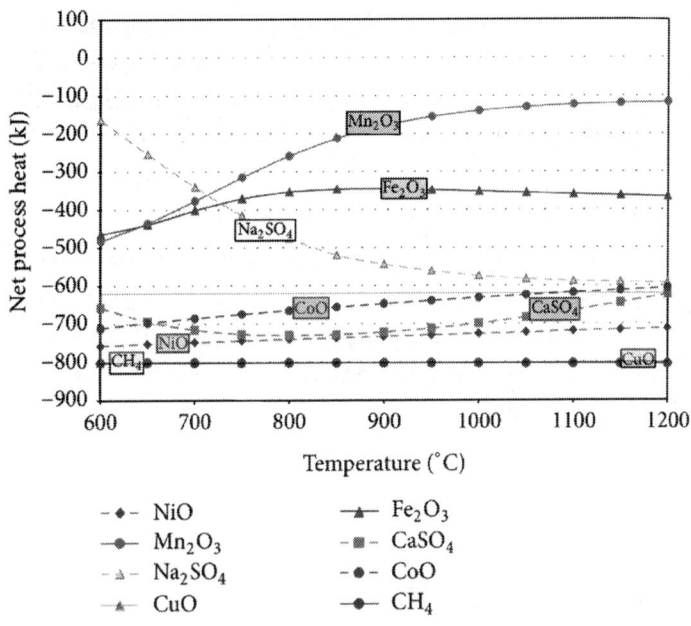

Figure 11: Net process heat for CLC of methane.

Weight of Oxygen Carrier

The CLC system involves circulation of the hot oxygen carrier from the air reactor to fuel reactor for reduction. The reduced oxygen carrier is then sent back to the air reactor for oxidation. This continuous recirculation of the solid oxygen carrier requires considerable energy. Hence, an estimate of weight of the oxygen carrier is important for the process engineer to help choose the right oxygen carrier. Table 4 shows the oxygen carrier requirement per mole methane used in the CLC process. It was observed that 4 moles of CuO, NiO, and CoO were required per mole methane used in the CLC process; hence, their weight is much higher than the sulfate oxygen carriers. The CuO, NiO, and CoO oxygen carriers were the heaviest, followed by Fe_2O_3 and Mn_2O_3. The Ca and Na sulfates were found to have lowest weight for use in this study. The favorable oxygen carrier was the lowest amount of the required oxygen carrier (in weight), and the ranking based on this criterion was found to be $CaSO_4 > Na_2SO_4 > Mn_2O_3 > Fe_2O_3 > NiO > CoO > CuO$. However, the trend of net energy obtainable with stoichiometric amount of oxygen carrier was found to be different. Hence, this criterion was compared on the basis of maximum net energy obtained per gram of the oxygen carrier (Table 4) and the ranking based on this criterion was observed to follow the sequence: $CaSO_4$ (−5.3657 kJ, 800°C) > Na_2SO_4 (−4.1705 kJ, 1200°C) > NiO (−2.5334 kJ, 600°C) > CuO (−2.5196 kJ, 1150°C) > CoO (−2.3714 kJ, 600°C) > Mn_2O_3 (−2.2981 kJ, 600°C) > Fe_2O_3 (−2.1855 kJ, 600°C).

Table 4: Weight of oxygen carriers

Oxygen carrier	Molecular wt. of oxygen carrier (for 1 mole)	Moles of oxygen carrier	Oxygen carrier weight (gram)	(max) Net process heat (kJ/gram) of oxygen carrier
CuO	79.55	4	318.18	−2.5196
NiO	74.7	4	298.77	−2.5334
CoO	74.93	4	299.73	−2.3714
Fe_2O_3	159.69	1.33	212.38	−2.1855
Mn_2O_3	157.87	1.33	209.97	−2.2981
$CaSO_4$	136.14	1	136.14	−5.3657
Na_2SO_4	142.04	1	142.04	−4.1705

Air Requirement

The CLC of methane can be accomplished by using a variety of oxygen carriers. The reduction of oxygen carriers is different to different potentials depending on CLC conditions. The regeneration of the reduced oxygen carrier as well as carbon oxidation reaction requires air for regeneration. The (stiochometric) requirement of air moles in the air reactor per mole of methane used in CLC was calculated to understand the operating costs related to air input. Figure 12 shows the total air moles required for different oxygen carriers in CLC of methane process in the temperature range from 600°C to 1200°C. The oxygen carrier and carbon were oxidised by air in the air reactor. It was observed that total moles of air used decreased with increase in temperature for NiO, Mn_2O_3, and CoO; it decreased from 9.11 to 8.69 moles for NiO, decreased from 6.40 to 3.24 moles for Mn_2O_3, and decreased from 8.68 to 7.73 moles for CoO, while the total air moles used remained almost constant at 9.50 moles for CuO and it increased with increase in temperature for Na_2SO_4, that is, from 2.56 to 7.55 moles for Na_2SO_4, while for Fe_2O_3, it first decreased from 6.21 (600°C) to 5.35 moles (850°C) and then increased to 5.55 moles (1200°C). The total air moles used first increased till 850°C and then slightly decreased with increase in temperature for $CaSO_4$; that is, it increased from 7.93 to 8.81 moles and then slightly decreased to 8.04 moles. Selectivity of the oxygen carriers according to the lowest amount of air moles used in the process was as follows: Na_2SO_4 (2.56 moles, 600°C) > Mn_2O_3 (3.24 moles, 1200°C) > Fe_2O_3 (5.35 moles, 850°C) > CoO (7.73 moles, 1200°C) > $CaSO_4$ (7.93 moles, 600°C) > NiO (8.69 moles, 1200°C) > CuO (9.49 moles, 1200°C).

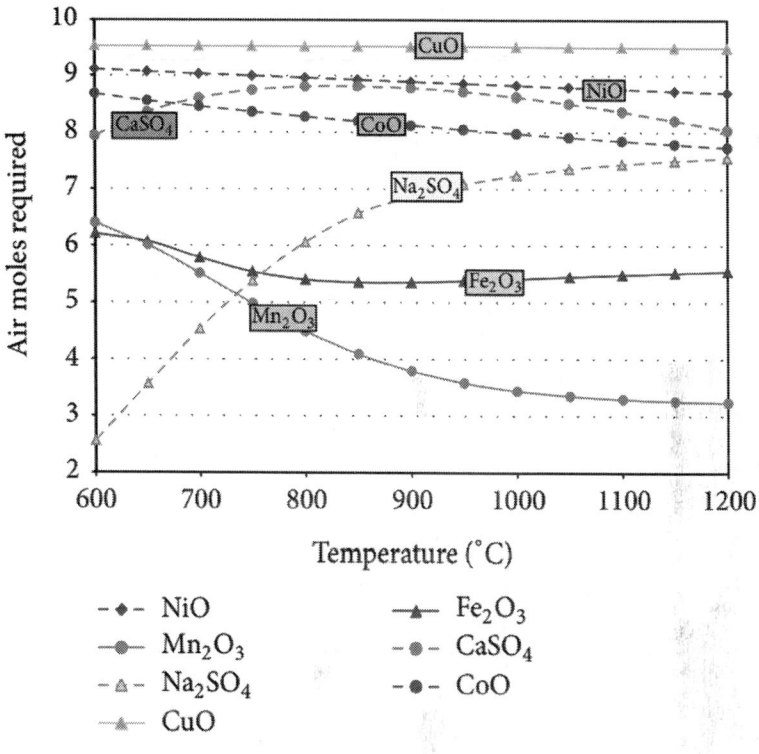

Figure 12: Air moles required for CLC of methane.

Effect of Pressure and Amount of Oxygen Carrier on Methane CLC Process

Pressure is an important process parameter. According to Le Chatelier's principle, the effect of increase in process pressure on the reaction in the CLC fuel reactor seems negative; that is, the H_2O and CO_2 yield decreased with increase in pressure at constant temperature for all oxygen carriers. The amount of oxygen carrier in the CLC system is based on the stoichiometric reaction. However, formation of syngas is seen due to thermodynamic limitations. An increase in the amount of oxygen carrier (greater than the stoichiometric amount) can enhance the conversion of fuel selectively to CO_2 and H_2O. However, this increase in the oxygen carrier quantity can put load on the solid conveying system of CLC. Table 5 shows the increase in the CO_2 yield by using different amounts (S—stoichiometric amount, 2S and 3S) of the oxygen carrier (NiO) at different temperatures. It can be seen that increase in the amount of oxygen carrier increases the solid input to the system in a huge quantity, but the CO_2 yield increases only meagrely. Hence, stoichiometric amount of the oxygen carrier is the optimum quantity to be used in the CLC system.

Table 5: Variation of CO_2 moles with increase in stoichiometric amount (S) of NiO

(a)

(S = stoichiometric amount of NiO)	1S	2S	3S
Weight (grams) of oxygen carrier (NiO)	318.2	636.4	954.6

(b)

Temperature (°C)	CO_2 moles			% increase (1S-2S)	% increase (1S–3S)
	(1S)	(2S)	(3S)		
600	0.9588	0.9984	0.9992	4.1335	4.2159
750	0.9353	0.9960	0.9980	6.4931	6.7037
1000	0.8696	0.9884	0.9941	13.6714	14.3269
1200	0.8151	0.9791	0.9893	20.1251	21.3765

Discussion

Properties of oxygen carrier significantly affect the overall performance of CLC process. The oxygen carriers selected for this study have been reported by researchers as potential oxygen carriers under experimental reaction and regeneration trials. The results obtained in this study inferred valuable information for selection of oxygen carrier for methane CLC. Table 6 shows the preferable oxygen carriers for the different criteria. From Table 6, it is observed that CuO and NiO showed better results among the oxide-based oxygen carriers. CuO and NiO have good desired product yield with negligible undesired product formation. Net process heat per gram of oxygen carrier is found to be maximum for NiO, followed by CuO. Among the sulphate-based oxygen carriers, $CaSO_4$ showed better results under most of the conditions considered. CLC operation involves the fluidization of oxygen carrier particles, and the energy required for solid oxygen carrier transport is a critical parameter for process operation. Hence, it is desirable to obtain maximum energy in CLC using lowest weight (amount) of oxygen carrier. Although the performance of oxygen carriers like NiO and $CaSO_4$ is similar, $CaSO_4$ produces higher net process heat in methane CLC per gram of oxygen carrier used in the process.

Table 6: Ranking of oxygen carrier

Sr. number	Criteria for section of oxygen carrier	Ranking of oxygen carrier
1	H_2O yield (moles)	$CuO > CaSO_4 > NiO > CoO > Na_2SO_4 > Fe_2O_3 > Mn_2O_3$
2	CO_2 yield (moles)	$CuO > NiO > CaSO_4 > CoO > Na_2SO_4 > Fe_2O_3 > Mn_2O_3$
3	$H_2O + CO_2$ yield (moles)	$CuO > CaSO_4 > NiO > CoO > Na_2SO_4 > Fe_2O_3 > Mn_2O_3$
4	Carbon formation (moles)	$Mn_2O_3 < Fe_2O_3 < CoO < NiO < Na_2SO_4 < CaSO_4 < CuO$
5	Syngas yield (moles)	$CuO > CaSO_4 > Na_2SO_4 > NiO > CoO > Mn_2O_3 > Fe_2O_3$
6	Net process heat (kJ)	$CuO > NiO > CaSO_4 > CoO > Na_2SO_4 > Mn_2O_3 > Fe_2O_3$
7	Oxygen carriers used (gram)	$CaSO_4 > Na_2SO_4 > Mn_2O_3 > Fe_2O_3 > NiO > CoO > CuO$

23. S. K. Sharma, J. S. Saini, I. M. Mishra, and M. P. Sharma, "Mirabilis leaves—a potential source of methane," Biomass, vol. 13, no. 1, pp. 13–24, 1987

24. S. K. Sharma, I. M. Mishra, M. P. Sharma, and J. S. Saini, "Effect of particle size on biogas generation from biomass residues," Biomass, vol. 17, no. 4, pp. 251–263, 1988

25. J. Huang and R. J. Crookes, "Assessment of simulated biogas as a fuel for the spark ignition engine,"Fuel, vol. 77, no. 15, pp. 1793–1801, 1998

26. P. Lunghi, R. Bove, and U. Desideri, "Life-cycle-assessment of fuel-cells-based landfill-gas energy conversion technologies," Journal of Power Sources, vol. 131, no. 1-2, pp. 120–126, 2004

27. R. Bove and P. Lunghi, "Electric power generation from landfill gas using traditional and innovative technologies," Energy Conversion and Management, vol. 47, no. 11-12, pp. 1391–1401, 2006

28. R. J. Spiegel, J. L. Preston, and J. C. Trocciola, "Fuel cell operation on landfill gas at Penrose Power Station," Energy, vol. 24, no. 8, pp. 723–742, 1999

29. R. J. Spiegel and J. L. Preston, "Technical assessment of fuel cell operation on landfill gas at the Groton, CT, landfill," Energy, vol. 28, no. 5, pp. 397–409, 2003

30. R. J. Spiegel, J. C. Trocciola, and J. L. Preston, "Test results for fuel-cell operation on landfill gas,"Energy, vol. 22, no. 8, pp. 777–786, 1997

31. W. T. Tsai, "Bioenergy from landfill gas (LFG) in Taiwan," Renewable and Sustainable Energy Reviews, vol. 11, no. 2, pp. 331–344, 2007

32. K. A. Kvenvolden, "A review of the geochemistry of methane in natural gas hydrate," Organic Geochemistry, vol. 23, no. 11-12, pp. 997–1008, 1995

33. S.-Y. Lee and G. D. Holder, "Methane hydrates potential as a future energy source," Fuel Processing Technology, vol. 71, no. 1–3, pp. 181–186, 2001.

34. E. Desa, "Submarine methane hydrates—potential fuel resource of the 21st century," Proceedings of Andra Pradesh Akademi of Sciences, vol. 5, no. 2, pp. 101–114, 2001

35. W. Rice, "Hydrogen production from methane hydrate with sequestering of carbon dioxide,"International Journal of Hydrogen Energy, vol. 31, no. 14, pp. 1955–1963, 2006

36. W. D. Gunter, T. Gentzis, B. A. Rottenfusser, and R. J. H. Richardson, "Deep coalbed methane in Alberta, Canada: a fuel resource with the potential of zero greenhouse gas emissions," Energy Conversion and Management, vol. 38, no. 1, pp. S217–S222, 1997

37. J. J. Lay, Y. Y. Li, T. Noike, J. Endo, and S. Ishimoto, "Analysis of environmental factors affecting methane production from high-solids organic waste," Water Science and Technology, vol. 36, no. 6-7, pp. 493–500, 1997

38. D. P. Chynoweth, J. M. Owens, and R. Legrand, "Renewable methane from anaerobic digestion of biomass," Renewable Energy, vol. 22, no. 1–3, pp. 1–8, 2001

39. Y. Li, S. Y. Park, and J. Zhu, "Solid-state anaerobic digestion for methane production from organic waste," Renewable and Sustainable Energy Reviews, vol. 15, no. 1, pp. 821–826, 2011

40. M. Hammad, D. Badarneh, and K. Tahboub, "Evaluating variable organic waste to produce methane,"Energy Conversion and Management, vol. 40, no. 13, pp. 1463–1475, 1999

41. G. J. MacDonald, "The future of methane as an energy resource," Annual Review of Energy, vol. 15, pp. 53–83, 1990

42. T. Mattisson, M. Johansson, and A. Lyngfelt, "The use of NiO as an oxygen carrier in chemical-looping combustion," Fuel, vol. 85, no. 5-6, pp. 736–747, 2006

43. L. Shen, M. Zheng, J. Xiao, and R. Xiao, "A mechanistic investigation of a calcium-based oxygen carrier for chemical looping combustion," Combustion and Flame, vol. 154, no. 3, pp. 489–506, 2008.

134 Energy Technology and Directions for the Future

134 Energy Technology and Directions for the Future

134 Energy Technology and Directions for the Future

134 Energy Technology and Directions for the Future

134 Energy Technology and Directions for the Future

44. T. Mattisson and A. Lyngfelt, "Capture of CO_2 using chemical-looping combustion," in Proceedings of the 1st Biennial Meeting of the Scandinavian-Nordic Section of the Combustion Institute, Göteborg, Sweden, April 2001.

45. T. Mattisson, E. Jerndal, C. Linderholm, and A. Lyngfelt, "Reactivity of a spray-dried $NiO/NiAl_2O_4$ oxygen carrier for chemical-looping combustion," Chemical Engineering Science, vol. 66, no. 20, pp. 4636–4644, 2011

46. B. M. Corbella and J. M. Palacios, "Titania-supported iron oxide as oxygen carrier for chemical-looping combustion of methane," Fuel, vol. 86, no. 1-2, pp. 113–122, 2007

47. M. Johansson, T. Mattisson, and A. Lyngfelt, "Investigation of Mn_3O_4 with stabilized ZrO2 for chemical-looping combustion," Chemical Engineering Research and Design, vol. 84, no. 9, pp. 807–818, 2006.

48. M. Rydén, E. Cleverstam, A. Lyngfelt, and T. Mattisson, "Waste products from the steel industry with NiO as additive as oxygen carrier for chemical-looping combustion," International Journal of Greenhouse Gas Control, vol. 3, no. 6, pp. 693–703, 2009.

49. C. R. Forero, P. Gayán, F. García-Labiano, L. F. de Diego, A. Abad, and J. Adánez, "Effect of gas composition in Chemical-Looping Combustion with copper-based oxygen carriers: fate of sulphur,"International Journal of Greenhouse Gas Control, vol. 4, no. 5, pp. 762–770, 2010.

50. R. Kuusik, A. Trikkel, A. Lyngfelt, and T. Mattisson, "High temperature behavior of NiO-based oxygen carriers for chemical looping combustion," Energy Procedia, vol. 1, pp. 3885–3892, 2009.

51. Q. Zafar, A. Abad, T. Mattisson, B. Gevert, and M. Strand, "Reduction and oxidation kinetics of $Mn_3O_4/Mg-ZrO_2$ oxygen carrier particles for chemical-looping combustion," Chemical Engineering Science, vol. 62, pp. 6556–66567, 2007.

52. A. Abad, J. Adánez, F. García-Labiano, L. F. de Diego, and P. Gayán, "Modeling of the chemical-looping combustion of

methane using a Cu-based oxygen-carrier," Combustion and Flame, vol. 157, no. 3, pp. 602–615, 2010.

53. Q. Song, R. Xiao, Z. Deng et al., "Chemical-looping combustion of methane with $CaSO_4$ oxygen carrier in a fixed bed reactor," Energy Conversion and Management, vol. 49, no. 11, pp. 3178–3187, 2008.

54. G. R. Kale, "Q science connect," pp. 1–12, 2012.

55. G. R. Kale, B. D. Kulkarni, and K. V. Bharadwaj, "Chemical looping reforming of ethanol for syngas generation: a theoretical investigation," International Journal of Energy Research, vol. 37, no. 6, pp. 645–656, 2013.

56. R. H. Perry and D. W. Green, Perry's Chemical Engineers' Handbook, McGraw-Hill, New York, NY, USA, 7th edition, 1997.

57. J. P. E. Cleeton, C. D. Bohn, C. R. Müller, J. S. Dennis, and S. A. Scott, "Clean hydrogen production and electricity from coal via chemical looping: identifying a suitable operating regime," International Journal of Hydrogen Energy, vol. 34, no. 1, pp. 1–12, 2009.

58. W. Xiang, S. Chen, Z. Xue, and X. Sun, "Investigation of coal gasification hydrogen and electricity co-production plant with three-reactors chemical looping process," International Journal of Hydrogen Energy, vol. 35, no. 16, pp. 8580–8591, 2010.

59. M. Ortiz, A. Abad, L. F. de Diego, F. García-Labiano, P. Gayán, and J. Adánez, "Optimization of hydrogen production by Chemical-Looping auto-thermal Reforming working with Ni-based oxygen-carriers," International Journal of Hydrogen Energy, vol. 36, no. 16, pp. 9663–9672, 2011.

60. HSC Chemistry [software]. Version 5.1 Pori: Outokumpu Research Oy, 2002.

61. W. R. Smith, "Computer software reviews, HSC chemistry for windows, 2. 0.," Journal of Chemical Information and Computer Sciences, vol. 36, pp. 151–152, 1996.

62. G. R. Kale, B. D. Kulkarni, and A. R. Joshi, "Thermodynamic study of combining chemical looping combustion and combined reforming of propane," Fuel, vol. 89, no. 10, pp. 3141–3146, 2010.

63. G. R. Kale and B. D. Kulkarni, "Thermodynamic analysis of dry autothermal reforming of glycerol," Fuel Processing Technology, vol. 91, no. 5, pp. 520–530, 2010.

64. G. R. Kale and B. D. Kulkarni, "An alternative process for gasoline fuel processors," International Journal of Hydrogen Energy, vol. 36, no. 3, pp. 2118–2127, 2011.

65. E. Jerndal, T. Mattisson, and A. Lyngfelt, "Thermal analysis of chemical-looping combustion," Chemical Engineering Research and Design, vol. 84, no. 9, pp. 795–806, 2006.

66. H.-B. Zhao, L.-M. Liu, D. Xu, C.-G. Zheng, G.-J. Liu, and L.-L. Jiang, "NiO/NiAl$_2$O$_4$ oxygen carriers prepared by sol-gel for chemical-looping combustion fueled by gas," Journal of Fuel Chemistry and Technology, vol. 36, no. 3, pp. 261–266, 2008.

67. H. Leion, A. Lyngfelt, and T. Mattisson, "Solid fuels in chemical-looping combustion using a NiO-based oxygen carrier," Chemical Engineering Research and Design, vol. 87, no. 11, pp. 1543–1550, 2009.

68. C. Dueso, A. Abad, F. García-Labiano et al., "Reactivity of a NiO/Al$_2$O$_3$ oxygen carrier prepared by impregnation for chemical-looping combustion," Fuel, vol. 89, no. 11, pp. 3399–3409, 2010.

69. C. Saha and S. Bhattacharya, "Comparison of CuO and NiO as oxygen carrier in chemical looping combustion of a Victorian brown coal," International Journal of Hydrogen Energy, vol. 36, no. 18, pp. 12048–12057, 2011.

70. T. Mattisson, H. Leion, and A. Lyngfelt, "Chemical-looping with oxygen uncoupling using CuO/ZrO$_2$ with petroleum coke," Fuel, vol. 88, no. 4, pp. 683–690, 2009.

71. K. Svoboda, A. Siewiorek, D. Baxter, J. Rogut, and M. Poho elý, "Thermodynamic possibilities and constraints for pure

hydrogen production by a nickel and cobalt-based chemical looping process at lower temperatures," Energy Conversion and Management, vol. 49, no. 2, pp. 221–231, 2008.

72. H. Jin, T. Okamoto, and M. Ishida, "Development of a novel chemical-looping combustion: synthesis of a looping material with a double metal oxide of CoO-NiO," Energy and Fuels, vol. 12, no. 6, pp. 1272–1277, 1998.

73. N. Ding, Y. Zheng, C. Luo, Q.-L. Wu, P.-F. Fu, and C.-G. Zheng, "Investigation into compound $CaSO_4$oxygen carrier for chemical-looping combustion," Journal of Fuel Chemistry and Technology, vol. 39, no. 3, pp. 167–168, 2011.

74. M. Zheng, L. Shen, and J. Xiao, "Reduction of $CaSO_4$ oxygen carrier with coal in chemical-looping combustion: effects of temperature and gasification intermediate," International Journal of Greenhouse Gas Control, vol. 4, no. 5, pp. 716–728, 2010.

75. Z. Deng, R. Xiao, B. Jin, and Q. Song, "Numerical simulation of chemical looping combustion process with $CaSO_4$ oxygen carrier," International Journal of Greenhouse Gas Control, vol. 3, no. 4, pp. 368–375, 2009.

76. L. Shen, M. Zheng, J. Xiao, and R. Xiao, "A mechanistic investigation of a calcium-based oxygen carrier for chemical looping combustion," Combustion and Flame, vol. 154, no. 3, pp. 489–506, 2008.

77. Z.-P. Gao, L.-H. Shen, J. Xiao, M. Zheng, and J.-H. Wu, "Analysis of reactivity of Fe-based oxygen carrier with coal during chemical-looping combustion," Journal of Fuel Chemistry and Technology, vol. 37, no. 5, pp. 513–520, 2009.

78. B. Wang, H. Lv, H. Zhao, and C. Zheng, "Experimental and simulated investigation of chemical looping combustion of coal with Fe_2O_3 based oxygen carrier," Procedia Engineering, vol. 16, pp. 390–395, 2011.

79. B.-W. Wang, R. Yan, Y. Zheng, H.-B. Zhao, and C.-G. Zheng, "Simulated investigation of chemical looping combustion

with coal-derived syngas and $CaSO_4$ oxygen carrier," Journal of Fuel Chemistry and Technology, vol. 39, no. 4, pp. 251–257, 2011.

80.   R. Xiao and Q. Song, "Characterization and kinetics of reduction of $CaSO_4$ with carbon monoxide for chemical-looping combustion," Combustion and Flame, vol. 158, no. 12, pp. 2524–2539, 2011.

81.   A. Shulman, E. Cleverstam, T. Mattisson, and A. Lyngfelt, "Chemical—looping with oxygen uncoupling using Mn/Mg-based oxygen carriers—oxygen release and reactivity with methane," Fuel, vol. 90, no. 3, pp. 941–950, 2011.

82.   L. F. de Diegoa, P. Gayán, J. Celaya, J. M. Palacios, and J. Adánez, "Operation of a 10 kWth chemical-looping combustor during 200 h with a $CuO–Al_2O_3$ oxygen carrier," Fuel, vol. 86, pp. 1036–1045, 2007.

83.   C. Saha, S. Zhang, K. Hein, R. Xiao, and S. Bhattacharya, "Chemical looping combustion (CLC) of two Victorian brown coals—part 1: assessment of interaction between CuO and minerals inherent in coals during single cycle experiment," Fuel, vol. 104, pp. 262–274, 2013.

Chapter 3

Photocatalytic Based Degradation Processes of Lignin Derivatives

Colin Awungacha Lekelefac[1], Nadine Busse[1],
Michael Herrenbauer[2], and Peter Czermak[1, 3, 4]

[1]Institute of Bioprocess Engineering and Pharmaceutical Technology, University of Applied Sciences Mittelhessen, 35390 Giessen, Germany

[2]Media University, Packaging Technology, 70569 Stuttgart, Germany

[3]Department of Chemical Engineering, Faculty of Engineering, Kansas State University, Manhattan, KS 66506, USA

[4]Faculty of Biology and Chemistry, Justus-Liebig-University Giessen, 35392 Giessen, Germany

In what follows, equations summarizing the formation of radical species under photocatalytic conditions shall be described. S stands for the lignin substrate while TiO_2 (h^+_{VB}) and TiO_2 (e^-_{CB}) represent the electron deficient (valence band) and electron-rich (conduction band) parts in the structure of TiO_2, respectively.

The initial photocatalytic process involves the generation of electron-hole pair in the semiconductor particles as a result of UV radiation [38, 39]. Figure 4 shows the excitation of an electron from the valence band to the conduction band initiated by light absorption with energy equal to or greater than the band gap of the semiconductor. This is expressed by (1).

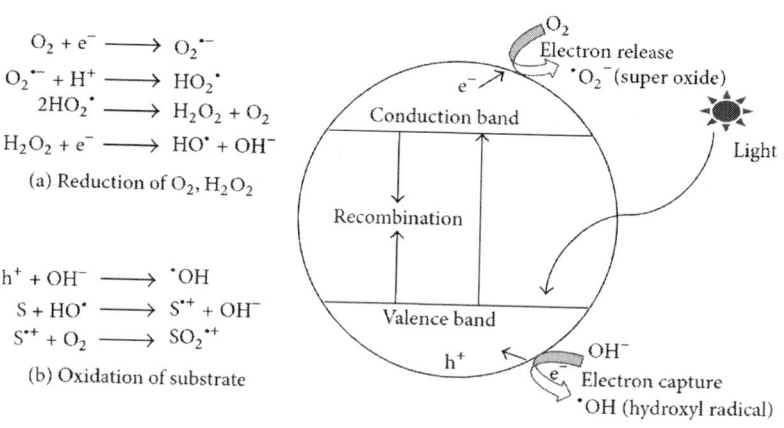

$$O_2 + e^- \longrightarrow O_2^{\bullet-}$$
$$O_2^{\bullet-} + H^+ \longrightarrow HO_2^{\bullet}$$
$$2HO_2^{\bullet} \longrightarrow H_2O_2 + O_2$$
$$H_2O_2 + e^- \longrightarrow HO^{\bullet} + OH^-$$

(a) Reduction of O_2, H_2O_2

$$h^+ + OH^- \longrightarrow {}^{\bullet}OH$$
$$S + HO^{\bullet} \longrightarrow S^{\bullet+} + OH^-$$
$$S^{\bullet+} + O_2 \longrightarrow SO_2^{\bullet+}$$

(b) Oxidation of substrate

Figure 4: Photocatalysis principle, adapted from Linsebigler et al. [23].

Upon excitation, the fate of the separated electron and hole can follow several pathways. Electron or holes can then react with hydroxyl ions (OH^-) or H_2O producing hydroxyl radicals ($^{\bullet}OH$) as shown in (2). Jaeger and Bard [40], Matthews [41], and Machado et al. [42] report that $^{\bullet}OH$ is the main oxidizing agent in the photocatalytic oxidation because of the unpaired electrons. Therefore, it can react fast and unspecifically with almost all organic compounds (S) [43] abstracting an electron with the formation of a radical organic species as shown in (3) [44]. Formation of singlet oxygen, hydroxyl, and superoxide radicals as principal reactive species in a photocatalytic process [38, 39] is as follows:

$$TiO_2 \xrightarrow{h\nu} TiO_2 \left(\begin{array}{c} e^-_{CB} \\ h^+_{VB} \end{array} \right) \tag{1}$$

$$TiO_2 (h^+) + H_2O \longrightarrow TiO_2 + HO^\bullet + H^+ \tag{2}$$

$$S + HO^\bullet \longrightarrow S^{\bullet+} + OH^- \tag{3}$$

$$S^{\bullet+} + O_2 \longrightarrow SO_2^{\bullet+} \tag{4}$$

$$TiO_2 (h^+) + S \longrightarrow TiO_2 + {}^\bullet S^+ \tag{5}$$

$$TiO_2 (e^-) + O_2 \longrightarrow TiO_2 + {}^\bullet O_2^- \tag{6}$$

$$HO^\bullet + {}^\bullet O_2^- \longrightarrow OH^- + {}^1O_2 \tag{7}$$

$$TiO_2 (e^-) + TiO_2 (h^+) \longrightarrow heat \tag{8}$$

The organic radicals and radical cations can, for example, react with molecular oxygen, to form organic peroxy radicals and peroxy radical cations respectively (4). The holes can oxidize organic compounds by electron abstraction to form organic cationic radicals (5) [42]. Superoxides can be formed by the reaction of electrons with electron acceptors such as O_2 (6). Meanwhile the formation of singlet oxygen can be from the reaction of hydroxyl radical and superoxide (7) [42]. Moreover, there is a possibility that electrons and holes recombine if electron acceptors are limited. In this case, recombination can take place in the volume of the semiconductor particle. When recombination takes place, radiation energy is lost or converted into heat (8) [45]. From investigations caried out by Mazellier et al. [46], (photochemistry of 2,6-dimethylphenol), it was postulated that hydrogen can be abstracted by α-carbonyl groups. In the same context, lignin derivatives having similar functionality can follow a similar pathway. In addition, oxidative chain reactions with the participation of ground-state oxygen can be initiated

INFLUENCE OF PROCESS PARAMETERS IN LIGNIN DEGRADATION

Varying process parameter could either have a positive or negative impact on the photocatalytic efficiency. The basic process parameters, such as catalyst concentration [4, 48, 56], substrate concentration [48, 51], addition of metal ion to TiO_2 catalyst [17, 50, 56–58], pH [47, 48], illumination [33, 48, 49, 59], and their influence shall be discussed in this subchapter. Table 2 gives an overview about starting reaction conditions and catalyst applied by some work groups while Table 3 portrays parameters, analytical methods, and results obtained. It is worthwhile noting that comparing the different photochemical processes poses a big challenge because of the wide variables involved. These discrepancies start already from the source and type of lignin followed by the differences in reactor design, illumination source, intensity of radiation, and different types of TiO_2 catalyst such as Fischer scientific rutile TiO_2 [49], TiO_2 (TiO_2—TP-2 of Fujititan), just to name a few.

Table 2: Summary of starting conditions for the photocatalytic degradation of lignin

Reference	Lignin source	Catalyst	Reaction conditions
MacHado et al. [42]	Peroxyformic acid lignins from eucalyptus grandis wood (EL1) EL1 + sodium borohydride	TiO$_2$ and H$_2$O$_2$	Lignin concentration: 0.25 mg/mL, T=25°C, pH 11, UV Vis radiation λ=300 nm, 400 W mercury lamp, cylindrical Pyrex glass reactor, constant oxygen bubbling
Ksibi et al. [47]	Water soluble lignin obtained from black liquor	TiO$_2$-P25 (Degussa)	Lignin concentration: 90 mg/L, T=20°C, pH 8.2, UV-radiation λ=290nm, pyrex reactor open to air, Philips HPK 125 W lamp
Kansal et al. [48]	Lignin from wheat straw kraft digestion	TiO$_2$-P25 (Degussa) and ZnO	Lignin concentration: 10–100 mg/L in 100 mL, ZnO catalyst dose (0.5–2.0 g/L), pH of the solution (pH 3–11), solar illumination, oxidant concentration: 3.06×10^{-6} M to 15.3×10^{-6} M, thin bed film slurry pond reactor, oxidant: sodium hypochlorite solution (4% available Cl$_2$)
Dahm and Lucia [49]	Whitewater from industrial process water that exits in paper machines	Fischer Scientific rutile TiO$_2$	Lignin concentration: 40 mg/L in 500 mL batch reactor, T=21°C–42°C, Rayonet photochemical chamber, 16 VWR 8-W black light, phosphor (350-nm) lamps, constant oxygen bubbling, power of illumination: 128 to 64 W, light intensity: 223–445 mW/cm^3

Reference	Lignin source	Photocatalyst	Experimental conditions
Portjanskaja and Preis [50]	Lignin, purchased from Aldrich	TiO_2 P25-N (Degussa)	Lignin concentration: 100 mg/L, pH 8, batch reactor system, reactor open to air, Phillips TLD 15 W/05 low-pressure luminescent mercury UV-lamp, UV-radiation $\lambda > 360$ nm, power density of irradiation $= 0.7$ mW/cm², visible light source
Tanaka et al. [51]	Lignin from coniferous wood	TiO_2 (TiO_2—TP-2 of Fujitan)	Lignin concentration: 0.003 to 0.03%, UV-radiation $\lambda > 310$ nm, cylindrical reaction vessel
Tonucci et al. [33]	Ca^{2+} and NH_4^+ lignin derivatives	TiO_2 as Degussa P25 + polyoxometalates (POM), H_2O_2	Open quartz tubes (20 mL) reactor, Y=20°C, 1 atm, multirays of ten UV lamps of 15 W power each, UV-radiation $\lambda > 254$ nm
Miyata et al. [24]	Picea glehnii wood flour	TiO_2/polyethylene oxide (PEO)	Reactor: open petri dish, T=30°C, t=48 h, 400 W mercury lamp
Shende et al. [17]	Kraft lignin	TiO_2-ZnO-ZrO_2	Lignin concentration: 1 mg/mL in 10 mL, 250 W Xenon lamp and AM 1.5 G lamp filter, power density of irradiation: 100 mW/cm², solar lamp simulator

Tian et al. [52] and Pan et al. [53]	Kraft lignin from black liquor	Ta_2O_5-IrO_2 and PbO_2 thin film TiO_2 nanotube/PbO_2	Lignin concentration: 30% (w/w), UV-radiation $\lambda > 365$ nm, t=10 min, power density of irradiation: 20 mW/cm², blue wave TM50 AS UV spot lamp, EG&G 2273 potentiostat/galvanostat to apply current, Ti/TiO_2NT/PbO_2 electrode as working electrode and Pt coil as outer electrode and Ag/AgCl as reference electrode
Awungacha Lekelefac et al. [54], [55]	Lignin sulfonate from paper waste water	TiO_2 as Degussa P25 TiO_2 from sol-gel process of $TiOSO_4$, TTIP	Lignin concentration: 500 mg/L in 200 mL, Osram Planon light source, irradiance: 30–40 W/m², UV-radiation λ: 280–420 nm, t=20 h, reactor open to air, recirculation system, flow rate 22.5 mL/min

Table 3: Parameters, analytical methods, and results from different work groups

Reference	Parameter studied	Analytics	Result
MacHado et al. [42]	Role of hydroxyl radicals, irradiation of lignin in the absence and presence of photocatalyst TiO_2 and H_2O_2	Ultraviolet-visible (UV-Vis) spectroscopy, ionization absorption spectroscopy (IAS), size exclusion chromatography (SEC)	A sharp decrease in the phenolic content observed for reactions involving direct photolysis; SEC: a reduction of almost 50% in the average molecular weight of lignin equal to 1.4 kD after 90 min of irradiation
Ksibi et al. [47]	Irradiation of lignin in the absence and presence of photocatalyst TiO_2-P25	UV-Vis spectroscopy, ^{13}C-nuclear magnetic resonance (NMR) solid state, total ion gas chromatography (TIC), induction coupling plasma (ICP), chemical oxygen demand (COD)	56% degradation rate with TiO_2 catalyst after 420 min, reaction time; ethyl acetate-extractable products showed vanillin, vanillic acid, palmitic acid, biphenyl structures, and 3,4,5-trimethoxy benzaldehyde; presence of magnesium and calcium ions; COD removal is higher for the initially low concentrations of lignin solution
Kansal et al. [48]	Catalyst dose, pH, oxidant concentration, initial substrate concentration, ZnO catalyst in slurry and immobilized mode	UV-Vis spectroscopy, COD	Optimum catalyst dose is 1 g/L; optimum oxidant concentration: 2×10^{-6} M; gradual decrease of absorption peak indicating decomposition of organics; COD removal is higher for the initially low concentrations of lignin solution

Dahm and Lucia [49]	Catalyst dose, illumination intensity	UV-Vis spectroscopy, total organic carbon (TOC), capillary ion electrophoresis analysis (CIA)	Gradual decrease in absorption peak indicating decomposition of organics; optimal catalyst dose of 10 mg/m; higher illumination intensities correlated well with higher initial degradation rate; 74% disappearance of TOC
Portjanskaja and Preis [50]	Influence of ferrous ions, N-doped catalyst effect, sprayed catalyst on support and submersed catalyst	COD, UV-Vis spectroscopy, biochemical oxygen demand (BOD), colorimetric measurement at 570 nm	Addition of Fe^{2+}, up to 2.8 mg/L leads to 25% increase in photocatalytic efficiency; sprayed catalyst exhibited 1.5 times higher efficiency than the one attached by submersion; negligible effect of N-doped catalyst; increase of aldehyde concentration over reaction time; neutral media was most beneficial for biodegradability; 80% of free phenols removed under neutral conditions

remains the most used catalyst as can be depicted from Table 2. TiO_2 is reported to be favored because of its nontoxic property, cost efficiency, and chemical and biological inertness. Moreover, TiO_2 possesses the most efficient photoactivity and the highest stability, thus making it suitable for industrial use [61].

ZnO has been described to degrade lignin under visible light sources [48, 62], whereas TiO_2 is mostly applied in connection to UV-light sources as highlighted in Table 2. However, both ZnO and TiO_2 possess energy band gap energy of 3.2 e.V. [23].

Additives such as SiO_2 [63], polyethylene oxide (PEO) [24], and polyethylene glycol [64] have been added to TiO_2 catalyst, particularly when applying immobilized catalyst. Addamo et al. [63] noted a high adhesion of TiO_2 to glass support material when precoating was done with SiO_2. Moreover, the precoating might have other advantages such as a hindering diffusion of Na^+ ions from the glass material into the nascent TiO_2 film during heat treatment processes. Analogous to the addition of SiO_2, polyethylene glycol (PEG) has also been introduced to mitigate catalyst surface activity, modify surface hydrophobicity, and also reduce agglomeration tendency of the TiO_2 gel or TiO_2 particles in the suspensions [64]. By a proper surface modification, interaction between catalyst and substrate can be enhanced [55, 63].

Kansal et al. [48] varied catalyst (ZnO) dose from 0.5 g/L to 2.0 g/L for 0.1 g/L kraft lignin solutions and found out that there was an optimum catalyst threshold value at 1 g/L which gives a catalyst to substrate ratio of 1 : 10.

Dahm and Lucia [49] examined catalyst dose from $2 \cdot 10^{-3}$ g/L to $1.2 \cdot 10^{-2}$ g/L for lignin solutions (from white water liner mill) of $4 \cdot 10^{-2}$ g/L (catalyst to lignin ratio: $5 \cdot 10^{-3}$–$3 \cdot 10^{-2}$) at pH 8 and obtained best energy efficiency values and lignin degradation rates with a catalyst loading of $1.0 \cdot 10^{-2}$ g/L.

In contrast, Ma et al. [56] applied far higher catalyst concentration compared to Dahm and Lucia [49]. Catalyst concentration was varied between 1 g/L and 10 g/L Pt/TiO_2. Best catalysis dose with respect to reaction turnover was obtained at 5 g/L Pt/TiO$_2$. With the increase of catalyst dose at pH 7, the reaction rate increased from 6.1 × 10^{-3} min^{-1} (1 g/L TiO$_2$) to 7·10^{-3} min^{-1} (5 g/L TiO$_2$) and 9.9·10^{-3} min^{-1} (10 g/L TiO$_2$).

Catalyst effect has been explained on the basis that optimum catalyst loading is dependent on the initial solute concentration. An increase in catalyst dosage leads to a corresponding increase of total active surface area for reactions [65]. To that, at higher TiO$_2$ concentrations, the photon flux is more easily intercepted by the catalyst before penetrating into the bulk of the system. At the same time, due to an increase in turbidity of the suspension with high dose of photocatalyst, there is a decrease in penetration of UV light and hence photoactivated volume of suspension or solution decreases [66].

In summary, authors have obtained best catalyst to lignin relations for different reaction designs and thus a general recommendation on catalyst dose is not possible. However, when lignin solution is treated with increasing catalyst loads, a corresponding increase in degradation rate is observed until a threshold value is reached [48–50, 56].

Influence of Metal Ion Addition (Doping) and Additives

The purpose of adding metal ion to photocatalyst is to mitigate band gap energy through the introduction of intraband gap states and as a consequence produce a bathochromic shift in the absorption spectrum [37]. Altering the absorption spectral range gives the possibility to exploit both the visible light spectrum and UV light sources. Metal ion doping is also introduced to serve as electron or hole traps in order to minimize recombination between generated electron-hole pairs [37].

Portjanskaja and Preis [50] studied the addition of Fe^{2+} ions to an acidic lignin solution and found an increase in photocatalytic oxidation (PCO) efficiency. The optimum Fe^{2+} ions quantity was

2.8 mg/L while using 100 mg/L lignin solution. Upon further elevation of Fe^{2+} ions concentration, a corresponding reduction of the photocatalytic oxidation efficiency of lignin was noted. Likewise, Ohnishi et al. [57] made a comparative study by doping platinum (Pt), silver (Ag), and gold (Au) ions to TiO_2. In these reactions, 50 mg of catalyst (TiO_2) was used with the addition of an equivalent 1.5 wt% (based on TiO_2) metal ion. The addition of noble metals brought about a faster decolorization of lignin. Au showed better results than Ag, followed by Pt. In the same context, adding sodium hypochlorite as oxidant to Pt/TiO_2 catalyst, an additional fivefold degradation rate was observed compared to that without doping [56]. Contradictory to the results described above, negligible effect of photocatalytic efficiency due to doping has been reported as well. Awungacha Lekelefac et al. [54] obtained little or no change in degradation rate by doping TiO_2-P25-SiO_2 catalyst with Pt ions (1 wt% relative to TiO_2 catalyst). Likewise Portjanskaja and Preis [50] noted a negligible change of photocatalytic efficiency of TiO_2 when doped with nitrogen.

Sarkanen et al. [67] and Gellerstedt and Lindfors [68] reported the bias of peroxides to oxidation with reagents such as permanganate to favor aromatic moieties. Oxidation agents like permanganate oxidizes predominantly aliphatic chains in alkaline and neutral media. However, by the application of H_2O_2 (Fenton system), lignin disappeared completely [33]. Tonucci et al. [33] conclude that, in order to satisfactorily conserve the organic material, the best compromise appears to be the TiO_2 photosystem, which shows low carbon consumption, good preservation of the aromatic rings, and greatly reduced mineralization.

In summary, different results have been obtained concerning the influence of noble metal ion addition. While some authors report an improvement in the photocatalytic efficiency upon their addition, others report their addition as having no considerable influence. However, for reactions in which an improvement in the photocatalytic efficiency was noticed, there was a threshold value to be considered. When the concentration of dopant surpasses this threshold value, electron-hole recombination is favored and this has a negative impact to photocatalysis. In such a case, the space-charge layer gets narrower and p-type dopants attract electrons and by virtue become negative. They would now then act as hole acceptor attracting holes. On the other hand, n-type dopants which act as electron donor centers and possess excess electrons attract holes as well [37].

Influence of Lignin Concentration

Once the initial lignin concentration becomes higher exceeding a threshold value, an inhibitory effect on the photodegradation was noted [47–49]. This threshold value varies and depends on the reaction system and reaction parameters such as optical density, catalyst concentration, and reaction volume. From the literature, different authors have implemented varying lignin concentrations probably to suit their reaction design. For example, Ksibi et al. [47] use 90 mg/L and Awungacha Lekelefac et al. [54] use 500 mg/L while Kansal et al. [48] applied 10 mg–100 mg/L.

Explanations arising from the findings are as follows: at low lignin concentrations, the incidental photonic flux irradiated interacts with the catalyst generating radicals (e.g., hydroxyl radicals (OH^-)) which allow a faster degradation [69]. On the other hand, high initial lignin concentrations may lead to tight adsorption which can suppress CO_2 evolution [51] and hence maintain chemical oxygen demand (COD) values. Moreover, low delignification yields may be due to an inhibitory effect because of autoxidation by low molecular weight lignin degradation products formed [24]. Also, due to the polymer structure of lignin which is cross-linked, this makes it difficult for radical species, acid, and the aldehyde compounds produced to spread into the inner region of the substrate hence limiting autoxidation. As a worst case, this might be the rate-determining step of delignification which is hindered [70–72].

In summary, it can be concluded that the time taken for complete degradation depends on the initial concentration of lignin and faster degradation occurs at low lignin concentrations.

Influence of pH

Varying pH entails an alteration in the properties of semiconductor-liquid interface [73], mainly related to the acid-base equilibrium of the adsorbed hydroxyl group [39]. Furthermore, pH also impacts lignin degradation rates [47, 48, 57]. In this context, several studies were carried out with partly contradictory outcomes.

Kansal et al. [48] made pH investigations 3–11 under solar light illumination using ZnO as catalyst. Maximum degradation was reached in alkaline conditions (pH 11). This is supported by Villaseñor and Mansilla [74] reporting an almost complete decolorization of kraft black liquor from pine wood at pH value of 11.6 in combination with ZnO catalyst. Similar results were achieved by Ohnishi et al. [57] with TiO_2 and ZnO being catalyst for bleaching alkaline lignin in aqueous solution with TiO_2 and ZnO being catalyst. High activities at neutral pH were also reported by Ohnishi et al. [57]. In contrast, Ma et al. [56] observed higher reaction rates and rapid degradation of a synthetic lignin wastewater (prepared by dissolving commercial lignin powder in aqueous solution; pH 11) in acidic solution (pH 3) than in alkaline solutions at pH 11, for either TiO_2 or Pt/TiO_2 catalysts.

Reconsidering the photocatalytic principle, the formed superoxide anion radicals ($^{\bullet}O_2^{-}$) are in a pH-dependent equilibrium with perhydroxyl radicals ($HO_2^{\bullet}$) as follows [75]:

$$HO_2^{\bullet} \rightleftharpoons H^+ + {}^{\bullet}O_2^{-} \quad pKa \sim 4.8$$

(9)

$^{\bullet}O_2^{-}$ undergoes dismutation reaction resulting in H_2O_2 and O_2 competing to any other $^{\bullet}O_2^{-}$ triggered reaction. In case of low pH operation conditions in aqueous solutions, $HO_2^{\bullet}$ becomes dominant whose reactivity is considerably higher compared to $^{\bullet}O_2^{-}$ [77]. Subsequently, $HO_2^{\bullet}$ initiates substrate (S) oxidation to the radical cation $(S^{\cdot+})$ and is itself reduced to H_2O_2 [78]. Thus, increased degradation rates can be reasonably expected supporting the results made by Ma et al. [56] in an acidic environment. $^{\cdot}O_2^{-}$ is extremely reactive in organic solvents [77]. Another aspect is the solubility of kraft lignin (soluble at pH > 10.5) which reduces with decreasing pH, whereas lignosulfonate should remain unaffected by pH in aqueous solution. Moreover, β–O–4 bonds have been described to be stable at acidic pH [11]. In fact, this could additionally explain the elevated degradation of kraft lignin made by Kansal et al. [48] and Villaseñor and Mansilla [74]. Nevertheless, the contradictory results gained by Ma et al. [56] still exist under the assumption that kraft lignin was used (which would be supported by the high pH of 11, obviously necessary for dissolving the lignin powder). Although most photocatalytic reactions described in the literature are in an aqueous milieu, lignin raw material and its fission products may however vary considerably. Therefore, the optimal pH is most likely to be reaction specific and has to be evaluated experimentally in principle.

Influence of Illumination

Many of the studies found in the literature so far have not dealt on this subject per se. What is found is the use of different illumination sources, each having a specified power and lamp type. However, all tend to emit UV-light between the range 280–420 nm. Table 2 depicts this in detail. Other illumination sources include the visible light spectrum.

In general terms, illumination influences in that it initiates photocatalysis by generating electron-hole pair in the semiconductor particles [38, 39]. Dahm and Lucia [49] altered illumination intensity while observing lignin degradation. In this study, 0.04 g/L lignin was used and light intensity was varied 223–445 mW/cm³. It was found out that higher illumination intensities correlated well with higher initial degradation rates and hence total lignin degradation [49]. Neppolian et al. [69] report degradation to be proportional to radiation intensity and best results are achieved for low lignin concentrations because of enhanced interaction between catalyst and incidental photonic flux.

In summary, high illumination power causes a corresponding high initial degradation rate at low lignin concentrations because maximum light penetration into the reaction medium is favored.

Process Analytical Methods

Various analytical techniques have been used to monitor lignin degradation. At the beginning of this subchapter, analytics revealing compounds formed from lignin degradation are treated. This includes, for example, gas chromatography (GC) and ^{1}H NMR (nuclear magnetic resonance). This is then followed by results qualitative analytic measurements such as ultraviolet-visible (UV-Vis) spectroscopy and dissolved carbon (DC). A list of authors, analytical techniques applied, and results achieved are outlined in Table 3.

Portjanskaja and Preis [50] studied lignin degradation by measuring the removal of phenols through colorimetric measurements. As a result of 24 h photocatalytic oxidation under neutral media conditions, 80% of free phenols were removed. Gas chromatography (GC) result from Ksibi et al. [47] attested vanillin, vanillic acid, palmitic acid, biphenyl, and 3,4,5-trimethoxy benzaldehyde structures after the photocatalysis of lignin from black liquor. This is in accordance with the findings of Tonucci et al. [33] reporting the formation of vanillin, hydroxyl methoxy-acetophenone, coniferyl alcohol, coniferyl aldehyde, methanol, formic acid, acetic acid, and small amounts of C-2 and C-3 alcohols as degradation products.

^{1}H NMR spectral analysis of lignin before illumination and after 24 h of illumination showing characteristic bands of aromatic rings, methoxy, and aliphatic side chains was compared with each other. Results revealed that the aromatic ring degraded faster than the aliphatic chain [51]. Fourier transformation infrared spectroscopy

(FTIR) showed bands corresponding to CH_3, CH_2, and CH which remained unchanged after illumination while bands corresponding to aromatic rings disappeared as a result of illumination [51, 53].

Results obtained from the combination of photochemical and electrochemical oxidation [52, 53] were similar to those of Tanaka et al. [51]. Here, ^{13}C-NMR confirmed the presence of the carbonyl functionality and the presence of vanillin and vanillic acid after 12 h photochemical-electrochemical oxidation. These results showed that the combination of a photocatalytic and an electrochemical oxidation significantly enhanced the efficiency of lignin degradation. This is because the applied anodic potential bias greatly suppressed the recombination of photogenerated electrons and holes [53].

Ultraviolet spectrophotometry offers a convenient method to qualitatively and quantitatively analyze lignin in solution [79]. This is reflected by the large number of publications using this technique [17, 33, 47–51, 56, 58,80]. This is most likely due to its simplicity to interpret lignin degradation. Lignins absorb UV light with high molar extinction coefficients because of the several methoxylated phenylpropane units of which they are composed [33]. Figure 8 depicts a series of photometric scans of ligninsulfonate from paper waste water showing a gradual reduction of absorbance during photocatalytic treatment [54]. Here, the absorption peaks are around 210 nm and 280 nm. Absorbance decreases with time implying the decomposition of lignin and the deterioration of chromophore groups [54].

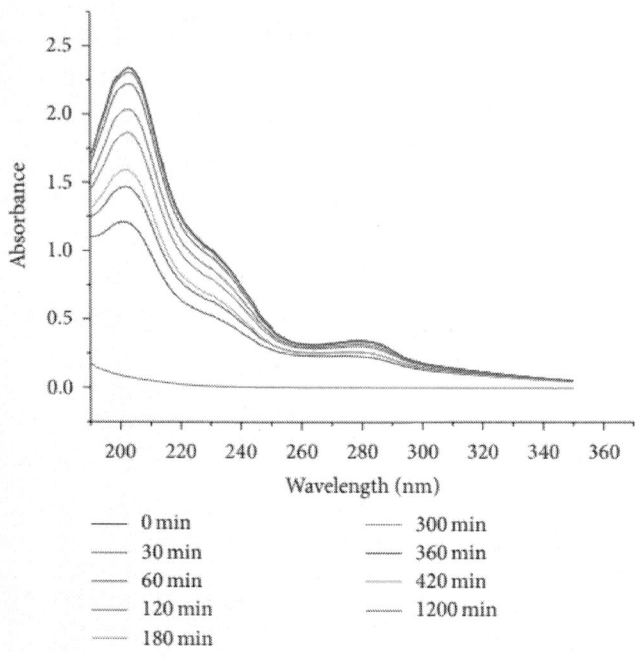

Figure 8: Time dependent UV-Vis absorption spectra of aqueous lignin solution from waste paper water irradiated with UV light (280–420 nm) for different time intervals. The spectra are obtained for sol-gel derived TiO$_2$ nanocrystalline coating (TiO$_2$-P25-SiO$_2$) [54].

Peaks at 210 nm correspond to portions of the unsaturated chains while those at 280 nm correspond to unconjugated phenolic hydroxyl groups [17] and the aromatic moiety [57] of the lignin molecule. The absorption tailing to the long wavelength region arises from the color of lignin [57]. Lignin degradation has been reported either at wavelength around 280 nm corresponding to unconjugated phenolic hydroxyl groups [17, 48, 51, 58] or for both wavelengths (210 nm and 280 nm) [33, 54, 57]. Kobayakawa et al. [81] noted some other absorbance at wavelengths lower than 250 nm and pointed out that this could be due to the modification of lignin fragmentations leading to the formation of transient species like methanol, ethanol, formaldehyde, formic acid, and oxalic acid among others.

Analytical methods to effectively quantify lignin degradation by calculating the oxygen demand by organic substances and remaining organic carbon before and after photocatalysis have been studied. Amongst the methods are dissolved carbon (DC) [49, 51, 58], chemical oxygen demand (COD) [47, 48, 50, 57, 80], biochemical oxygen demand (BOC) [50], dissolved organic carbon (DOC) [56], and American dye manufacture institute value (ADMI) [56]. COD and BOD describe the oxygen demand by organic substances to be converted to CO and CO_2 and H_2O and NH_3. Total organic carbon (TOC) describes the amount of carbon bound in an organic compound while DOC describes the dissolved fraction of organic carbon. ADMI measures the amount of dyestuff in water.

Decolorization of lignin solution has been reported to be another parameter observed during photocatalytic degradation. Color is an indirect indicator of the lignin amount. The higher the color intensity of the solution is, the greater the lignin content is (high concentrated lignin solutions, e.g., black liquor, appear dark brown) [82]. Thus, color changes can be interpreted as conversion of lignin to transient species or conversion to CO_2 and H_2O. Awungacha Lekelefac et al. [54] observed a gradual change from the characteristic yellow lignin (when highly diluted) to a colorless liquid after a period of 20 h with sol-gel derived TiO_2 nanocrystalline coatings on sintered borosilicate glass as depicted in Figure 9. A corresponding decrease in DC values close to 82% was observed for TiO_2-P25-SiO_2 catalyst confirming degradation. This is shown in Figure 10.

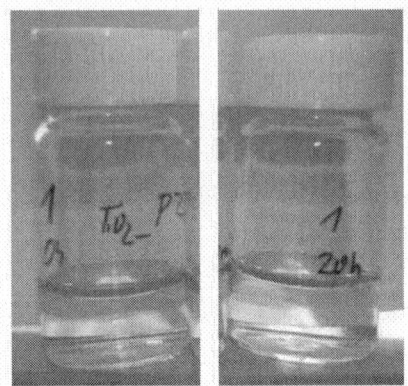

Figure 9: Gradual change from the characteristic yellow lignin sulfonate to a colorless liquid after a period of 20 h. Catalyst: TiO_2-P25 (Degussa) + TEOS, UV-light, 25°C, lignin concentration: 0.5 g/L [54].

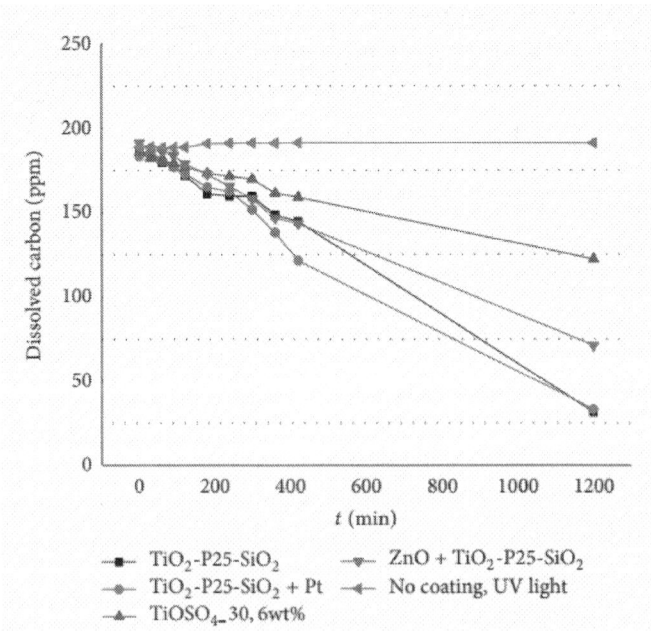

Figure 10: Variation of DC with time of aqueous lignin solution from waste paper water irradiated with UV light (280–420 nm) for different

time intervals. The spectra are obtained for sol-gel derived TiO_2 nanocrystalline coatings (TiO_2-P25-SiO_2 + Pt, TiO_2-P25-SiO_2, $TiOSO_4$−30.6 wt%, ZnO + TiO_2-P25-SiO_2) [54].

These findings are analogous to that of Ohnishi et al. [57] who reported the bleaching of lignin when illuminated continuously and that the solution becomes colorless. To that, the chemical oxygen demand (COD) value decreases, generating carbon dioxide and a small amount of carbon monoxide as the main gaseous products. COD removal was reported to be effective at low lignin concentrations as compared to high lignin concentrations [48].

Another applied analytical technique is fluorescence detection directly coupled to a high performance liquid chromatography (HPLC) as a means to identify nonaliphatic component in the complex mixture of lignin degradation products [54]. Fluorescence emission in lignin is attributed to aromatic structures such as conjugated carbonyl, biphenyl, phenylcoumarone, and stilbene groups [83, 84]. Awungacha Lekelefac et al. [54] observed peaks on both HPLC and fluorescence chromatograms suggesting the production of new substances and fluorophores.

Despite developed analytical technologies, analyzing lignin degradation products remains challenging. Proofs such as mass spectroscopy (MS), HPLC, ^{13}C, or ^{1}H-NMR spectra from photocatalytic lignin degradations are not yet established. The setback to qualitatively and quantitatively analyze lignin and its degradation products starts from the native lignin polymer itself with its indefinite polymeric structure and multiple bond types. Also, the influence of different pretreatments additives and the wide variety of compounds obtainable from its degradation makes lignin analysis challenging [85]. Moreover, lignin streams could contain proteins, inorganic salts, and other potential poisons that generally complicate catalysis [4].

The challenge to identify and separate the products streams derived from the photocatalytic degradation lignin is also worth noting. Lignin product stream is highly functionalized and conventional techniques such as gas chromatography have the disadvantage of requiring a time-consuming derivatization step.

Also, because of high boiling point of substances arising from lignin degradation, it is not easily applicable. High performance liquid chromatography (HPLC) seems to be the remedy because analysis can be carried out without derivatization but the exact identification of the separated substances is difficult because of the numerous peaks arising from such a chromatogram [87]. Unfortunately, well-established databases such as that of the national institute of standard and technology (NIST) [88] cannot give information on HPLC-MS chromatograms. This is because of the ionization sources such as electrospray ionization (ESI) and atmospheric pressure chemical ionization (APCI) for LC-MS while GC-MS is based on electron ionization. A remedy to this can be the development of a database of model lignin compounds based on a unanimous HPLC-MS measuring procedure. This would mean much time and cost expensive investments for adequate personnel and material.

Though product identification based on peak superposition gives information on a possible product, this cannot always be true for lignin degradation because of the large number of possible degradation products. In order to properly identify products from lignin degradation, product isolation techniques are to be implemented and further analytic methods such as tandem mass spectroscopy (MS/MS) and other advanced structure enhancing research techniques such as ^{1}H-NMR, ^{13}C-NMR have to be studied extensively.

PHOTOCATALYSIS AND ITS SUITABILITY AS INTEGRATED TECHNOLOGY IN MULTISTAGE CONCEPTS WITH BIOCATALYSIS

Depending on lignin treatment specification, photocatalysis will find application for complete lignin mineralization to CO_2 and H_2O (single stage) or as integrated technology in a multistage system aiming at partial conversion of lignin macromolecules to organic low molecular weight intermediates for a consecutive biological oxidation, finally generating value added end-products [15].

For biological oxidation of lignin or lignin derivatives, H_2O_2-dependent ligninolytic heme peroxidases (POXs, including lignin peroxidase (LiP, EC 1.11.1.14), manganese peroxidase (MnP, EC 1.11.1.13), versatile peroxidase (VP, EC 1.11.1.16), O_2-dependent laccases (Lac, EC 1.10.3.2), and extracellular enzymes fromBasidiomycetous white-rot fungi are the most efficient lignin degraders in nature [89]. Thus, such enzymatic systems, especially POXs (with high redox potentials), have attracted much interest as industrial biocatalyst [86, 90]. The enzyme degradation mechanism (in nature) is facilitated by nonenzymatic processes mainly through free $^\square$OH radicals also generated by the fungus (Fenton-type reaction). Those $^\square$OH radicals enable the required physical contact between the enzymes and structural units of the lignin molecule due to numerous nonspecific oxidative reactions [91]. Both the photocatalytic (equations (2)–(7)) and the POX reaction mechanism recently reviewed by Busse at al. [25] are quite similar. Consequently a combination of photocatalysis followed by an enzymatic oxidation maybe a promising concept utilizing lignin derivatives, for example, originated from industrial effluents [15, 92]. The advantages forming biobased products are as follows; see also Table 4 in this context. A reduction of the lignin polymerization degree via photocatalysis, at best, in more biodegradable intermediates and a simultaneous detoxification, causes savings in enzyme costs, since it will be expected that less enzyme loads are necessary for sufficiently rapid reaction rates [15, 91]. As a result, hydraulic retention times in bioreactor systems should be diminished. Moreover, the delignification is expected to be enhanced simultaneously, which was in recent years shown in a dual system by Kamwilaisak and Wright [93] using $TiO_2/H_2O_2/UV$ for photocatalytic pretreatment and Lac (from Trametes versicolor) in the subsequent biocatalytic step. Within 24 h, they obtained in their dual system an elevation in delignification of 20% (without H_2O_2) up to at least 50% when H_2O_2 was present.

Table 4: Main advantages and disadvantages, photocatalysis versus enzymatic biocatalysis

Lignin degradation process	Advantages	Disadvantages
Photocatalysis (single stage)	(i) Relatively fast degradation of complex as well as nonbiodegradable organic (macro)molecules (ii) Stable catalysts (iii) Moderate reaction conditions	(i) Not selective (ii) Limited selection of operating conditions (iii) Energy costs
Enzymatic conversion (single stage)	(i) More selective (ii) Moderate reaction conditions	(i) High enzyme loads required, otherwise lignin hydrolysis is quite slow [16] (ii) Enzymes are less stable[1] (iii) cost-intensive (POXs)[2]

[1] For example, POXs are sensitive to high H_2O_2 concentrations.

[2] According to Torres and Ayala [86].

In previous studies, the treatment of lignin (from the pulp and paper industry) containing effluents by fungus as a biological system (excreting appropriate lignin degrading enzyme cocktails) were more focused for posttreatment exclusively (Durán et al. [92], Reyes et al. [94], and González et al. [95]). Shende et al. [17] even examined ligninolytic bacteria, with photooxidized kraft lignin as substrate.

Although POXs are potential industrial biocatalysts [86], no application studies were found for the direct use in such dual systems as described above. The major reason may be the complexity of the reaction mechanism (inclusive lignin derivatives as substrate and their analysis) per se, on the one hand, slowing down research and development processes. On the other hand, POXs are sensitive to their cosubstrate H_2O_2, once it is supplied in excess causing considerable inactivation (for details, refer to Busse et al. [25]). At the present, several studies are carried out modifying these enzymes regarding enhanced stability, activity, and selectivity as well. Hence, it can be expected that their technological applicability will be raised significantly right after successful modification is reached [86].

CONCLUDING REMARKS

It is widely assumed that the photocatalytic degradation of lignin follows a radical reaction pathway which is similar to that considered in thermal, electrochemical, and biochemical processes. However, reporting on the degradation pathway of lignin derivatives and even that of lignin model compounds are still a major challenge. This is probably due to the complex nature and variety of possible degradation products. Indeed the mechanism is far more complex considering other factors such as type of lignin, type of catalyst, pH, illumination source, and additives.

Generally, comparing the different photochemical processes poses a big challenge because of the wide variables involved. These discrepancies start from the source and type of lignin followed by differences in reactor design, illumination source, intensity of radiation, and different types of catalyst. An idea would be to have a specific reference reaction with well-defined starting parameters which include lignin type, source, and purity, catalyst specifications, illumination source, and intensity so as to ease comparison of results.

Basic process parameters, such as catalyst concentration, substrate concentration, addition of metal ion to catalyst, pH, and illumination, have been discussed. Despite developed analytical technologies, analyzing lignin degradation products remains challenging. Proofs such as mass spectroscopy (MS), HPLC, ^{13}C, or ^{1}H-NMR spectra from photocatalytic lignin degradations are not yet established.

In order to properly identify products from lignin degradation, product isolation techniques are to be implemented and further analytic methods such as MS/MS and other advanced structure enhancing research techniques such as ^{1}H-NMR and ^{13}C-NMR have to be studied extensively. A remedy to this can be the development of a databank of model lignin compounds based on a unanimous HPLC-MS measuring procedure. This would mean much time and cost expensive investments for adequate personnel and material.

Photocatalysis is denoted as the most popular lignin pretreatment technology, besides ozonation [96]. Photocatalyzed lignin may be an appropriate substrate for a consecutive biocatalytic process using ligninolytic enzymes (POX and/or Lac) as supported by experimental results of Kamwilaisak and Wright [93]. Combining the advantages of both catalytic processes savings in the overall process costs will be expected in addition to elevated lignin conversion. Nonetheless, extensive research work including POX modifications is still required.

AUTHORS' CONTRIBUTION

Colin Awungacha Lekelefac and Nadine Busse contributed equally to this work.

ACKNOWLEDGMENTS

The authors gratefully thank the Federal Ministry of Education and Research (BMBF) for funding (FKZ17N0310). The researchers also thank the Hessen State Ministry of Higher Education, Research and Arts for the financial support within the Hessen initiative for scientific and economic excellence (LOEWE).

REFERENCES

1.   Crude Oil and Commodity Prices, http://www.oil-price.net/.

2.   Research, Directorate-General for World Energy, Technology and Climate Policy Outlook (WETO), Luxembourg Office for Official Publications of the European Communities, 2003.

3.   Green and Natural Polymers Are on the Rise, 2014, http://www.polymersolutions.com/blog/green-and-natural-polymers-on-the-rise/.

4.   J. Zakzeski, P. C. Bruijnincx, A. L. Jongerius, and B. M. Weckhuysen, "The catalytic valorization of lignin for the production of renewable chemicals," Chemical Reviews, vol. 110, no. 6, pp. 3552–3599, 2010.

5.   B. Kamm, M. Kamm, P. R. Gruber, and S. Kromus, Biorefineries-Industrial Processes and Products Status Quo and Future Directions, vol. 1, Wiley-VCH, 2006.

6.   R. L. Howard, E. Abotsi, E. L. J. van Rensburg, and S. Howard, "Lignocellulose biotechnology: issues of bioconversion and enzyme production," African Journal of Biotechnology, vol. 2, no. 12, pp. 602–619, 2003.

7.   M. Stöcker, "Bio- und BTL-Kraftstoffe in der Bioraffinerie: katalytische Umwandlung Lignocellulose-reicher Biomasse mit porösen Stoffen," Angewandte Chemie, vol. 120, no. 48, pp. 9340–9351, 2008.

8.   J.-P. Lange, "Lignocellulose conversion: an introduction to chemistry, process and economics," Biofuels, Bioproducts and Biorefining, vol. 1, no. 1, pp. 39–48, 2007. ·

9.   D. A. I. Goring, "The physical chemistry of lignin," Pure and Applied Chemistry, vol. 5, no. 1-2, pp. 233–310, 1962.

10.   M. Ek, G. Gellerstedt, and G. Henriksson, Wood Chemistry and Biotechnology: Pulp and Paper Chemistry and Technology, vol. 1, Walter de Gruyter GmbH & Co. KG, Berlin, Germany, 2009.

11. B. Saake and R. Lehnen, "Lignin," in Ullmann›s Encyclopedia of Industrial Chemistry, Wiley-VCH Verlag GmbH & Co. KGaA, Weinheim, Germany, 2012.

12. M. N. S. Kumar, A. K. Mohanty, L. Erickson, and M. Misra, "Lignin and its applications with polymers,"Journal of Biobased Materials and Bioenergy, vol. 3, no. 1, pp. 1–24, 2009.

13. R. J. A. Gosselink, E. de Jong, B. Guran, and A. Abächerli, "Co-ordination network for lignin—standardisation, production and applications adapted to market requirements (EUROLIGNIN),"Industrial Crops and Products, vol. 20, no. 2, pp. 121–129, 2004.

14. D. Fengel and G. Wegener, Wood: Chemistry, Ultrastructure, Reactions, Verlag Kessel, 1984.

15. D. Mantzavinos and E. Psillakis, "Enhancement of biodegradability of industrial wastewaters by chemical oxidation pre-treatment," Journal of Chemical Technology and Biotechnology, vol. 79, no. 5, pp. 431–454, 2004.

16. M. Dashtban, H. Schraft, and W. Qin, "Fungal bioconversion of lignocellulosic residues; Opportunities & perspectives," International Journal of Biological Sciences, vol. 5, no. 6, pp. 578–595, 2009.

17. A. Shende, R. Jaswal, D. Harder-Heinz, A. Menan, and R. Shende, "Intergrated photocatalytic and microbial degradation of kraft lignin," Cleantech, pp. 120–123, 2012.

18. F. S. Chakar and A. J. Ragauskas, "Review of current and future softwood kraft lignin process chemistry," Industrial Crops and Products, vol. 20, no. 2, pp. 131–141, 2004.

19. W. G. Glasser and H. R. Glasser, "Evaluation of lignin›s chemical structure by experimental and computer simulation techniques," Paperi ja Puu, vol. 63, pp. 71–83, 1981.

20. M. Erickson, S. Larsson, and G. E. Miksche, "Zur Struktur des Lignins der Fichte," Acta Chemica Scandinavica, vol. 27, pp. 903–914, 1973.

21. H. Nimz, "Das Lignin der Buche—Entwurf eines Konstitutionsschemas," Angewandte Chemie—International Edition, vol. 86, pp. 336–344, 1974.

22. D. V. Evtuguin, C. P. Neto, J. Rocha, and J. D. P. de Jesus, "Oxidative delignification in the presence of molybdovanadophosphate heteropolyanions: mechanism and kinetic studies," Applied Catalysis A: General, vol. 167, no. 1, pp. 123–139, 1998.

23. A. L. Linsebigler, G. Lu, and J. T. Yates Jr., "Photocatalysis on TiO$_2$ surfaces: Principles, mechanisms, and selected results," Chemical Reviews, vol. 95, no. 3, pp. 735–758, 1995.

24. Y. Miyata, K. Miyazaki, M. Miura, Y. Shimotori, M. Aoyama, and H. Nakatani, "Solventless delignification of wood flour with TiO$_2$/poly(ethylene oxide) photocatalyst system," Journal of Polymers and the Environment, vol. 21, no. 1, pp. 115–121, 2013. ·

25. N. Busse, D. Wagner, M. Kraume, and P. Czermak, "Reaction kinetics of versatile peroxidase for the degradation of lignin compounds," The American Journal of Biochemistry and Biotechnology, vol. 9, no. 4, pp. 365–394, 2013.

26. M. Tien and T. K. Kirk, "Lignin-degrading enzyme from phanerochaete chrysosporium: Purification, characterization, and catalytic properties of a unique H$_2$O$_2$-requiring oxygenase," Proceedings of the National Academy of Sciences, vol. 81, pp. 2280–2284, 1984.

27. T. K. Kirk, M. Tien, P. J. Kersten, M. D. Mozuch, and B. Kalyanaraman, "Ligninase of Phanerochaete chrysosporium. Mechanism of its degradation of the non-phenolic arylglycerol β-aryl ether substructure of lignin," Biochemical Journal, vol. 236, no. 1, pp. 279–287, 1986.

28. T. Lundell, R. Wever, R. Floris, et al., "Lignin peroxidase L3 from Phlebia radiata: pre-steady-state and steady-state studies with veratryl alcohol and a non-phenolic lignin model compound 1-(3,4-dimethoxyphenyl)-2-(2-methoxyphenoxy) propane-1,3-diol," European Journal of Biochemistry, vol. 211, no. 3, pp. 391–402, 1993.

29. H. E. Schoemaker, T. K. Lundell, A. I. Hatakka, and K. Piontek, "The oxidation of veratryl alcohol, dimeric lignin models and lignin by lignin peroxidase: the redox cycle revisited," FEMS Microbiology Reviews, vol. 13, no. 2-3, pp. 321–331, 1994.

30. J. M. Palmer, P. J. Harvey, and H. E. Schoemaker, "The role of peroxidases, radical cations and oxygen in the degradation of lignin [and discussion]," Philosophical Transactions of the Royal Society of London A, vol. 321, no. 1561, pp. 495–505, 1987.

31. C. Hill, Wood Modification. Chemical, Thermal, and Other Processes, John Wiley & Sons, Chichester, UK, 2006.

32. Y.-H. Xu, H.-R. Chen, Z.-X. Zeng, and B. Lei, "Investigation on mechanism of photocatalytic activity enhancement of nanometer cerium-doped titania," Applied Surface Science, vol. 252, no. 24, pp. 8565–8570, 2006.

33. L. Tonucci, F. Coccia, M. Bressan, and N. D›Alessandro, "Mild photocatalysed and catalysed green oxidation of lignin: a useful pathway to low-molecular-weight derivatives," Waste and Biomass Valorization, vol. 3, no. 2, pp. 165–174, 2012.

34. A. Castellan, N. Colombo, C. Vanucci, P. Fornier de, H. Violet, and H. Bouas-Laurent, "Photodegradation of lignin. A photochemical study of an O-methylated -carbonyl -1 lignin model dimer: 1,2-di(3'4'-dimethoxyphenyl) ethanone (deoxyveratroin)," Journal of Photochemistry and Photobiology A, vol. 51, pp. 451–467, 1990.

35. A. Fujishima, X. Zhang, and D. A. Tryk, "TiO_2 photocatalysis and related surface phenomena," Surface Science Reports, vol. 63, no. 12, pp. 515–582, 2008.

36. C. S. Turchi and D. F. Ollis, "Photocatalytic reactor design: an example of mass-transfer limitations with an immobilized catalyst," The Journal of Physical Chemistry, vol. 92, no. 23, pp. 6852–6853, 1988.

37. R. Subasri, M. Tripathi, K. Murugan, J. Revathi, G. V. N. Rao, and T. N. Rao, "Investigations on the photocatalytic activity of sol–gel derived plain and Fe^{3+}/Nb^{5+} doped titania coatings on

glass substrates," Materials Chemistry and Physics, vol. 124, pp. 63–68, 2010.

38. N. Serpone, "Relative photonic efficiencies and quantum yields in heterogeneous photocatalysis,"Journal of Photochemistry and Photobiology A: Chemistry, vol. 104, no. 1–3, pp. 1–12, 1997.

39. M. R. Hoffmann, S. T. Martin, W. Choi, and D. W. Bahnemann, "Environmental applications of semiconductor photocatalysis," Chemical Reviews, vol. 95, no. 1, pp. 69–96, 1995.

40. C. D. Jaeger and A. J. Bard, "Spin trapping and electron spin resonance detection of radical intermediates in the photodecomposition of water at titanium dioxide particulate systems," The Journal of Physical Chemistry, vol. 83, no. 24, pp. 3146–3152, 1979. ·

41. R. W. Matthews, "Hydroxylation reactions induced by near-ultraviolet photolysis of aqueous titanium dioxide suspensions," Journal of the Chemical Society, Faraday Transactions, vol. 80, no. 2, pp. 457–471, 1984.

42. A. E. H. Machado, A. M. Furuyama, S. Z. Falone, R. Ruggiero, D. D. S. Perez, and A. Castellan, "Photocatalytic degradation of lignin and lignin models, using titanium dioxide: the role of the hydroxyl radical," Chemosphere, vol. 40, no. 1, pp. 115–124, 2000. ·

43. W. Sigg and L. Stumm, Aquatische Chemie. Stuttgart. 2 Auflage, B.G. Teubner, Stuttgart, Germany, 1991.

44. O. Legrini, E. Oliveros, and A. M. Braun, "Photochemical processes for water treatment," Chemical Reviews, vol. 93, no. 2, pp. 671–698, 1993.

45. G. Rothenberger, J. Moser, M. Grätzel, N. Serpone, and D. Sharma, "Charge carrier trapping and recombination dynamics in small semiconductor particles," Journal of the American Chemical Society, vol. 107, no. 26, pp. 8054–8059, 1985.

46. P. Mazellier, M. Sarakha, A. Rossi, and M. Bolte, "The aqueous photochemistry of 2,6-dimethylphenol. Evidence for the fragmentation of the C-C bond," Journal of Photochemistry and Photobiology A, vol. 115, no. 2, pp. 117–121, 1998.

47. M. Ksibi, S. B. Amor, S. Cherif, E. Elaloui, A. Houas, and M. Elaloui, "Photodegradation of lignin from black liquor using a UV/TiO$_2$ system," Journal of Photochemistry and Photobiology A, vol. 154, no. 2-3, pp. 211–218, 2003.

48. S. K. Kansal, M. Singh, and D. Sud, "Studies on TiO$_2$/ZnO photocatalysed degradation of lignin," Journal of Hazardous Materials, vol. 153, no. 1-2, pp. 412–417, 2008.

49. A. Dahm and L. A. Lucia, "Titanium dioxide catalyzed photodegradation of lignin in industrial effluents," Industrial and Engineering Chemistry Research, vol. 43, no. 25, pp. 7996–8000, 2004.

50. E. Portjanskaja and S. Preis, "Aqueous photocatalytic oxidation of lignin: the influence of mineral admixtures," International Journal of Photoenergy, vol. 2007, Article ID 76730, 7 pages, 2007.

51. K. Tanaka, R. C. R. Calanag, and T. Hisanaga, "Photocatalyzed degradation of lignin on TiO$_2$," Journal of Molecular Catalysis A: Chemical, vol. 138, no. 2-3, pp. 287–294, 1999.

52. M. Tian, J. Wen, D. MacDonald, R. M. Asmussen, and A. Chen, "A novel approach for lignin modification and degradation," Electrochemistry Communications, vol. 12, no. 4, pp. 527–530, 2010.·

53. K. Pan, M. Tian, Z.-H. Jiang, B. Kjartanson, and A. Chen, "Electrochemical oxidation of lignin at lead dioxide nanoparticles photoelectrodeposited on TiO$_2$ nanotube arrays," Electrochimica Acta, vol. 60, pp. 147–153, 2012.

54. C. Awungacha Lekelefac, J. Hild, P. Czermak, and M. Herrenbauer, "Photocatalytic active coatings for lignin degradation in a continuous packed bed reactor," International Journal of Photoenergy, vol. 2014, Article ID 502326, 10 pages, 2014.

55. C. Awungacha Lekelefac, P. Czermak, and M. Herrenbauer, "Evaluation of photocatalytic active coatings on sintered glass tubes by methylene blue," International Journal of Photoenergy, vol. 2013, Article ID 614567, 9 pages, 2013.

56. Y.-S. Ma, C.-N. Chang, Y.-P. Chiang, H.-F. Sung, and A. C. Chao, "Photocatalytic degradation of lignin using Pt/TiO$_2$ as the catalyst," Chemosphere, vol. 71, no. 5, pp. 998–1004, 2008.

57. H. Ohnishi, M. Matsumura, H. Tsubomura, and M. Iwasaki, "Bleaching of lignin solution by a photocatalyzed reaction on semiconductor photocatalysts," Industrial and Engineering Chemistry Research®, vol. 28, no. 6, pp. 719–724, 1989.

58. C. A. K. Gouvêa, F. Wypych, S. G. Moraes, N. Durán, and P. Peralta-Zamora, "Semiconductor-assisted photodegradation of lignin, dye, and kraft effluent by Ag-doped ZnO," Chemosphere, vol. 40, no. 4, pp. 427–432, 2000.

59. A. V. Vähätalo, K. Salonen, M. Salkinoja-Salonen, and A. Hatakka, "Photochemical mineralization of synthetic lignin in lake water indicates enhanced turnover of aromatic organic matter under solar radiation," Biodegradation, vol. 10, no. 6, pp. 415–420, 1999.

60. H. de Lasa, B. Serrano, and M. Salaices, Photocatalytic Reaction Engineering, Springer, New York, NY, USA, 2005.

61. K. Hashimoto, H. Irie, and A. Fujishima, "TiO 2 photocatalysis: A historical overview and future prospects," Japanese Journal of Applied Physics, vol. 44, no. 12, pp. 8269–8285, 2005.

62. M. A. Behnajady, N. Modirshahla, and R. Hamzavi, "Kinetic study on photocatalytic degradation of C.I. Acid Yellow 23 by ZnO photocatalyst," Journal of Hazardous Materials, vol. 133, no. 1–3, pp. 226–232, 2006.

63. M. Addamo, V. Augugliaro, A. di Paola et al., "Photocatalytic thin films of TiO$_2$ formed by a sol-gel process using titanium tetraisopropoxide as the precursor," Thin Solid Films, vol. 516, no. 12, pp. 3802–3807, 2008.

64. N. Negishi, K. Takeuchi, and T. Ibusuki, "Preparation of the TiO$_2$ thin film photocatalyst by the dip-coating process," Journal of Sol-Gel Science and Technology, vol. 13, no. 1–3, pp. 691–694, 1998.

65. L. Rideh, A. Wehrer, D. Ronze, and A. Zoulalian, "Photocatalytic degradation of 2-chlorophenol in TiO$_2$ aqueous suspension: modeling of reaction rate," Industrial and Engineering Chemistry Research, vol. 36, no. 11, pp. 4712–4718, 1997.

66. R.-A. Doong, C.-H. Chen, R. A. Maithreepala, and S.-M. Chang, "The influence of pH and cadmium sulfide on the photocatalytic degradation of 2-chlorophenol in titanium dioxide suspensions," Water Research, vol. 35, no. 12, pp. 2873–2880, 2001.

67. K. V. Sarkanen, A. Islam, and C. D. Anderson, "Ozonation," in Methods in Lignin Chemistry, S. Y. Lin and C. W. Dence, Eds., Springer Series in Wood Science, pp. 387–406, Springer, Berlin, Germany, 1992.

68. G. Gellerstedt and E.-L. Lindfors, "Structural changes in lignin during kraft pulping," Holzforschung, vol. 38, no. 3, pp. 151–158, 1984.

69. B. Neppolian, H. C. Choi, M. V. Shankar, B. Arabindoo, and V. Murugesan, in Proceedings of the International Symposium on Environmental Pollution Control and Waste Management (EPCOWM ‹02), p. 647, 2002.

70. C. Pouteau, P. Dole, B. Cathala, L. Averous, and N. Boquillon, "Antioxidant properties of lignin in polypropylene," Polymer Degradation and Stability, vol. 81, no. 1, pp. 9–18, 2003.

71. J. Hafrén, T. Fujino, and T. Itoh, "Changes in cell wall architecture of differentiating tracheids of Pinus thunbergii during lignification," Plant and Cell Physiology, vol. 40, no. 5, pp. 532–541, 1999.

72. T. Dizhbite, G. Telysheva, V. Jurkjane, and U. Viesturs, "Characterization of the radical scavenging activity of lignins natural antioxidants," Bioresource Technology, vol. 95, no. 3, pp. 309–317, 2004.

73. K. Hofstadler, R. Bauer, S. Novalic, and S. G. Heisier, "New reactor design for photocatalytic wastewater treatment with TiO_2 immobilized on fused-silica glass fibers: photomineralization of 4-chlorophenol,"Environmental Science & Technology, vol. 28, no. 4, pp. 670–674, 1994.

74. J. Villaseñor and H. D. Mansilla, "Effect of temperature on kraft black liquor degradation by ZnO-photoassisted catalysis," Journal of Photochemistry and Photobiology A: Chemistry, vol. 93, no. 2-3, pp. 205–209, 1996.

75. B. H. Bielski, D. E. Cabelli, L. A. Ravindra, and A. B. Ross, "Reactivity of HO_2/O2- Radicals in aqueous solution," Journal of Physical Chemistry, vol. 14, pp. 1041–1100, 1985.

76. B. H. J. Bielski and A. O. Allen, "Mechanism of the disproportionation of superoxide radicals," Journal of Physical Chemistry, vol. 81, no. 11, pp. 1048–1050, 1977. ·

77. B. Halliwell and J. M. C. Gutteridge, "The importance of free radicals and catalytic metal ions in human diseases," Molecular Aspects of Medicine, vol. 8, no. 2, pp. 89–193, 1985.

78. J. M. Palmer, P. J. Harvey, and H. E. Schoemaker, "The role of peroxidases, radical cations and oxygen in the degradation of lignin [and discussion]," Philosophical Transaction of the Royal Society A, vol. 321, no. 1561, pp. 495–505, 1987.

79. S. Y. Yin and C. W. Dense, Methods in Lignin Chemistry, Springer, New York, NY, USA, 1992.

80. P. Kumar, S. Kumar, and N. K. Bhardwaj, "Photocatalytic oxidation of elemental chlorine free bleaching effluent with UV/TiO_2," in Proceedings of the 2nd International Conference on Environmental Science and Technology (ICEST '11), Singapore, February 2011.

81. K. Kobayakawa, Y. Sato, S. Nakamura, and A. Fujishima, "Photodecomposition of Kraft lignin catalyzed by titanium dioxide," Bulletin of the Chemical Society of Japan, vol. 62, no. 11, pp. 3433–3436, 1989.·

82. R. B. Kinstre, "An overview of strategies for reducing the environmental impact of bleach-plant effluents," Tappi Journal, vol. 76, no. 5, pp. 105–113, 1993.

83. A. Castellan, H. Choudhury, R. Stephen Davidson, and S. Grelier, "Comparative study of stone-ground wood pulp and native wood 3. Application of fluorescence spectroscopy to a study of the weathering of stone-ground pulp and native wood," Journal of Photochemistry and Photobiology, A: Chemistry, vol. 81, no. 2, pp. 123–130, 1994.

84. B. Albinsson, S. Li, K. Lundquist, and R. Stomberg, "The origin of lignin fluorescence," Journal of Molecular Structure, vol. 508, no. 1–3, pp. 19–27, 1999. ·

85. S. Baumberger, A. Abaecherli, M. Fasching et al., "Molar mass determination of lignins by size-exclusion chromatography: towards standardisation of the method," Holzforschung, vol. 61, no. 4, pp. 459–468, 2007.

86. E. Torres and M. Ayala, Biocatalysis Based on Heme Peroxidases, Springer, New York, NY, USA, 1st edition, 2010.

87. R. Pecina, P. Burtscher, G. Bonn, and O. Bobleter, "GC-MS and HPLC analyses of lignin degradation products in biomass hydrolyzates," Fresenius Zeitschrift für Analytische Chemie, vol. 325, no. 5, pp. 461–465, 1986.

88. NIST, NIST Standard Reference Database, The National Institute of Standards and Technology, 2014,http://www.nist.gov/srd/nist1a.cfm.

89. T. K. Kirk and R. L. Farrell, "Enzymatic "combustion": the microbial degradation of lignin," Annual Review of Microbiology, vol. 41, pp. 465–501, 1987.

90. A. T. Martinez, "High redox potential peroxidases," in Industrial Enzymes, Structure, Function and Applications, K.-B. Becker, Ed., pp. 477–488, Springer, Amsterdam, The Netherlands, 1st edition, 2007.

91. M. Dashtban, H. Schraft, T. A. Syed, and W. Qin, "Fungal biodegradation and enzymatic modification of lignin,"

International Journal of Biochemistry and Molecular Biology, vol. 1, no. 1, pp. 36–50, 2010.·

92. N. Durán, E. Esposito, L. H. Innocentini-Mei, and V. P. Canhos, "A new alternative process for Kraft E1 effluent treatment," Biodegradation, vol. 5, no. 1, pp. 13–19, 1994.

93. K. Kamwilaisak and P. C. Wright, "Investigating laccase and titanium dioxide for lignin degradation,"Energy and Fuels, vol. 26, no. 4, pp. 2400–2406, 2012.

94. J. Reyes, M. Dezotti, H. Mansilla, J. Villaseñor, E. Esposito, and N. Durán, "Biomass photochemistry-XXII: combined photochemical and biological process for treatment of Kraft E1 effluent," Applied Catalysis B: Environmental, vol. 15, no. 3-4, pp. 211–219, 1998.

95. L. F. González, V. Sarria, and O. F. Sánchez, "Degradation of chlorophenols by sequential biological-advanced oxidative process using Trametespubescens and TiO_2/UV," Bioresource Technology, vol. 101, no. 10, pp. 3493–3499, 2010.

96. J. H. Lora and W. G. Glasser, "Recent industrial applications of lignin: a sustainable alternative to nonrenewable materials," Journal of Polymers and the Environment, vol. 10, no. 1-2, pp. 39–48, 2002.·

Four Main Objectives for the Future of Chemical and Process Engineering Mainly Concerned by the Science and Technologies of New Materials Production

Jean-Claude Charpentier

President of the European Federation of Chemical Engineering, Department of Chemical Engineering/CNRS, Ecole Supérieure de Chimie, Physique et Electronique de Lyon, BP 2077, 69616 Villeurbanne Cedex, France

ABSTRACT

Today the chemical and process engineering especially involving chemical reactor engineering has to answer to the changing needs of the chemical and related process industries such as petroleum, petrochemical, bituminous, pharmaceutical and health, agro and food, environment, iron and steel, building materials, paints, glass, surfactants, electronics, cosmetic and perfume, etc., and to meet market demands. So being a key to survival in globalisation of trade and competition, the evolution of chemical engineering is thus necessary. And to satisfy both, the market requirements for specific end-use properties of the products manufactured in (bio) chemical reactors and the social and the resource-saving and environmental constraints of the industrial-scale processes and technologies, it is shown that a necessary progress is coming via a multidisciplinary and time and length multiscale approach. In such a frame the future for the science and technologies of new materials can be summarized by four main objectives:

(1)   a total multiscale control of the process (or the procedure) to increase selectivity and productivity, i.e., nanotailoring of materials with controlled structure;

(2)   a design of novel equipment based on scientific principles and new operation modes and methods of production: process intensification;

(3)   product design and engineering: manufacturing end-use properties with a special emphasis on complex fluids and solids technology;

(4)   an implementation of the multiscale and multidisciplinary computational chemical engineering modelling and simulation to real-life situations: from the molecule to the overall complex production scale into the entire production site.

Moreover, chemical and process engineering will also be increasingly involved and concerned with the application of life cycle assessment to new material design and production and its use but also to the plant and the equipment together with the associated services.

INTRODUCTION: THE MOVING WORLD NECESSARY REQUIREMENTS FOR CHEMICAL AND RELATED INDUSTRIES

The world moves forward. For the developing and industrializing countries, there is low cost of manpower and less constraining local production regulations. For the industrialised countries, there is a rapid development in consumer demand and constraints stemming from public concern over questions of environment and safety. In response to these changes, the world of chemistry and related industry including process industries such as petroleum, petrochemical, bituminous, pharmaceutical and health, agro and food, environment, textile, iron and steel, building materials, glass, surfactants, cosmetic and perfume, electronics, are confronted, from the technological and scientific point of view with a double challenge:

- To research innovative processes for the production of commodity and intermediate products. By no longer selecting processes only on the basis of economic exploitation but by seeking compensating gains resulting from the increased selectivity and savings linked to the process itself. This requires valorization of safety, health and environmental aspects, including the value of non-polluting technologies, reduction of raw material and energy losses and product and by-product recyclability as well. The industry will have to process with large plants supplying bulk products in large volumes. The customer will buy a process which is non-polluting, defect-free and perfectly safe.

- To progress from the traditional intermediate chemistry to new specialities and active material chemistry and related industries. This concerns industries involved with food products, with products for human, animal and vegetal health, with advanced materials, along with the chemistry/

biology interface, i.e., the postgenomic world involving proteomics and metabolomics. This concerns also upgrading and conversion of petroleum feed stocks and intermediates, conversion of coal derived chemicals or synthesis gas into fuels, hydrocarbons or oxygenates. The aim is characterized by new market objectives, with sales and competitiveness dominated by the end-use properties of a product linked to its quality or shape and size (i.e., chemical and biological stability, degradability, chemical, biological and therapeutic activity, aptitude to dissolution, mechanical, rheological, electrical, thermal, optical, magnetic characteristics for solids and solid particles together with size, shape, colour, touch, cohesion, friability, rugosity, tasks, succulence, aesthetics, sensory properties, etc.). Control of the end-use property and expertise in the design of the process, its permanent adjustments to variety and changing demand along with speed in reacting to market conditions will be the dominant elements. Indeed for these new specialities and active materials the consumer buys the product which is the most efficient and the first on the market. He will have to pay high prices and expect a large benefit from these short life time and high-margin products.

Being a key to survival in global markets including the previous needs and challenges, chemical and process engineering necessitates the today evolution in teaching and revolution in research. Indeed the objective of petroleum engineering, then of chemical engineering broadened to process engineering, is the synthesis, design, scale-up or scale-down, operation, control and optimization of industrial processes that change the state, microstructure and (bio-agro)chemical composition of material through physic (bio)chemical separations (distillation, absorption, extraction, drying, filtration, agitation, precipitation, fluidization, emulsification, crystallization, agglomeration, etc.) as well as chemical, catalytic, biochemical, electrochemical, photochemical or agrochemical reactions. It involves the whole of scientific and technical knowledge necessary for physico-chemical and biological transformations of raw material and energy into the targeted products necessitated by the customer. But it is important to note that today 60% of all products that a chemical company sells to its client are crystalline, polymeric or amorphous solids. These products need to have a clearly defined physical shape or texture in order to meet the designed and desired quality standards. This also applies to paste like and emulsified products. Instead of classical basic and industrial chemicals, new developments increasingly concern highly targeted and specialized materials, active compounds and special effect chemicals. These are much more complex in terms of molecular structure than classical chemicals and require a global approach for that manufacturing.

LE GÉNIE DU TRIPLE "PROCESSUS–PRODUIT–PROCÉDÉ": THE INTEGRATED MULTIDISCIPLINARY AND MULTISCALE APPROACH OF CHEMICAL AND PROCESS ENGINEERING

Thus chemical and process engineering is now concerned with the understanding and development of systematic procedures for the design and optimal operation of chemical and process systems, ranging from microsystems to industrial-scale continuous and batch processes, as presented in Fig. 1 in using the concept of chemical supply chain [1]. This chain starts with chemical and other products that industry must size and characterize at the molecule level. Subsequent step aggregates the molecules into clusters, particles, and thin films as single or multiphase systems that finally take the form of macroscopic mixtures—solids, paste-like or emulsion products. Transition from chemistry or biology to engineering, one move to the design and analysis of the production units, which are integrated into a process that in turn becomes part of an industrial site with multiple processes. Finally this site is a part of the commercial enterprise driven by market considerations and demands involving product quality.

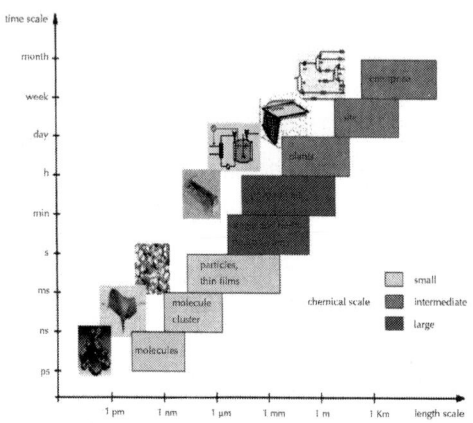

Figure 1: Chemical supply chain [1].

In this supply chain, it should be emphasized that product quality is determined at the micro- and nano-level and that a product with a desired property must be investigated for both structure and function. This involves a thorough understanding of the structure/property relationship at both molecular (e.g., surface physics and chemistry) and microscopic levels. And the ability to control microstructure formation to obtain the end-use properties of a fluid or solid product is the key to success and will help design and control product quality and make the leap from the nano-level to the process level.

This necessitates an integrated system approach for a multiscale and multidisciplinary modelling of the complex, simultaneous and often coupled momentum, heat and mass transfer phenomena and processes taking place at different time scales (10^{-15} to 10^8 s) and length scales (10^{-8} to 10^4 m) encountered in industrial practise (Fig. 2) involving approaches at the different length scales presented in Fig. 3.

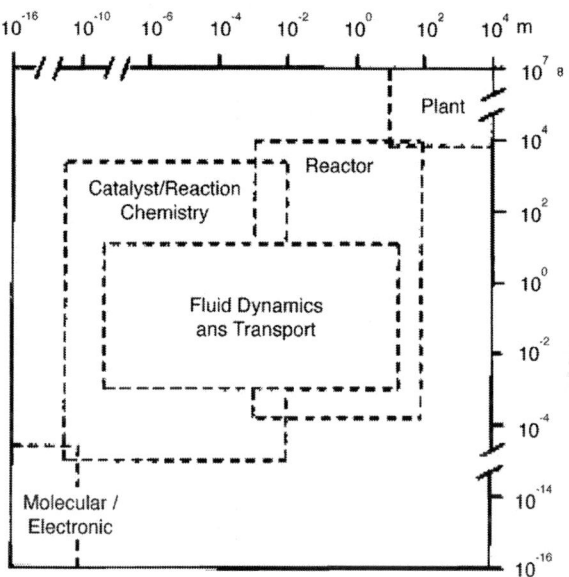

Figure 2: The length and time scales covered in the multiscale approach.

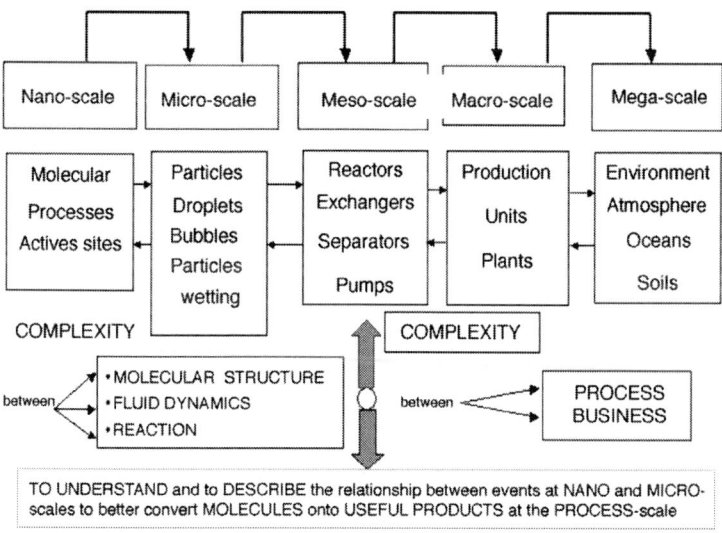

Figure 3: Organizing levels of increasing complexity underlie new view of chemical engineering.

So organizing scales and complexity levels in process engineering is now necessary to understand and to describe the relationships between events at nano- and microscales to better convert molecules into useful products at the process scale. And organizing levels of complexity by translating molecular processes (that I define by the name processus) into phenomenological macroscopic laws to create and to control the end-use properties and functionality of products manufactured by a continuous process underlie the today new views of chemical and process engineering. This can be defined by le Genie du triple "processus–produits–procédé" (the triplet molecular processes–product–process engineering, 3P engineering) with an integrated system approach of complex processes and phenomena occurring at different time and length scales [2].

This explains why, in addition to the basic notions of unit operations, coupled transfers and classical tools of chemical engineering, that is, in addition to the fundamentals of chemical engineering (separation engineering, chemical reaction engineering, catalysis, transport phenomena, optimization and process control), this integrated multidisciplinary and multiscale approach is a supplementary and considerable advantage for the development and the success of this engineering science in terms of concept and paradigms. This approach is of a great help in order to analyze, design and operate processes able to manufacture a product—first on the market—(often having a short life cycle) with the desired property and optimally thanks to processes involving if possible zero defect, zero pollution and zero accident.

And it will be possible to understand and to describe the relationship between events at nano-scale and micro-scale to better convert molecules into useful products at the zero pollution and zero accident process scale thanks to the large breakthroughs (a) in molecular modelling (both theory and computer simulation), (b) in scientific instrumentation and non-invasive measurement techniques (NMR, TAP, tomographic techniques, spectroscopic or monochromatic ellipsometry, i.e., diffusing wave spectroscopy, etc.) and related micro- and nanotechnologies in connection with image processing and (c) in powerful computational tools and capabilities, necessary for the treatment of generalized local information (increasing use of computational fluid dynamics or mixing, e.g. CFDLIB, FLUENT, PHOENICS, FLOW 3D, FIDAP, FLOWMAP, etc.) [2].

CHEMICAL AND PROCESS ENGINEERING: QUO VADIS?

The previous general considerations on the future of chemical engineering concern four main objectives.

Total Multiscale Control of the Process (or the Procedure) to Increase Selectivity and Productivity

This necessitates the "intensification" of operations and the use of precise nano- and micro-technology design. This is the case of molecular information engineering encountered for the supported organometallic catalysis or for supramolecular catalysis where instead of using porous support for heterogeneous catalyst, synthetic materials with targeted properties are now conceived and designed. Indeed, central to a successful catalytic process is the development of an effective catalyst which is a complex system in both composition and functionality. And the ability to better control its microstructure and chemistry allows for the systematic manipulation of the catalyst's activity, selectivity, and stability.

Nanotailoring of Materials with Controlled Structure: Opportunities for Molecular Engineering in Catalysis

Indeed through the control of pore opening and crystallite size and/or a proper manipulation of stoichiometry and component dispersion there exists now ability to engineer via nanostructure synthesis novel structures at the molecular and supramolecular levels, leading to the creation of nanoporous and nanocrystalline materials [3] (Fig. 4). These materials both possess an ultrahigh surface-to-volume ratio, which offers a greatly increased number of active sites for carrying out catalytic reactions.

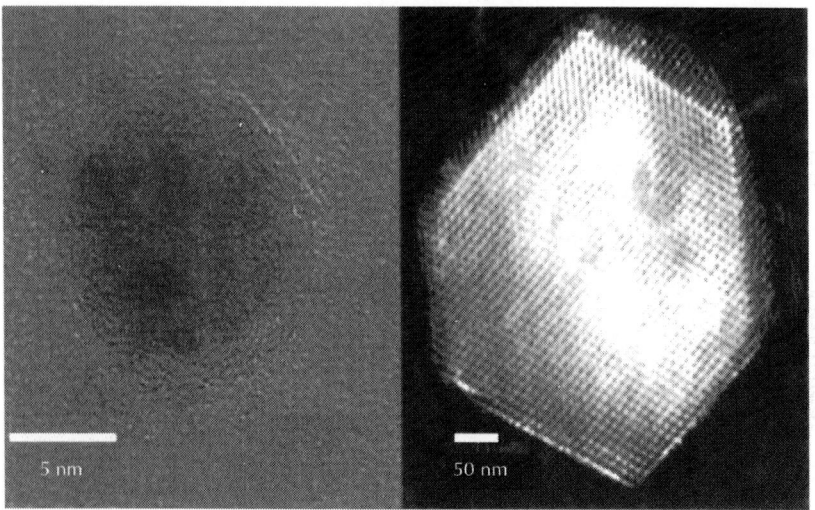

Figure 4: Tailoring of materials with controlled structure [3].

Nanocrystalline processing includes the tailoring of size-dependent electronic properties, homogeneous multicomponent systems, defect chemistry, and excellent phase dispersion. This provides nanocrystalline catalysts with greatly improved catalytic activity over conventional systems and multifunctionalities necessary for complex applications. For example for structure-sensitive reactions such as photocatalysis over titania used for decomposition of chemical wastes such as chloroform or trichloroethylene, catalytic activity depends not only on the number of active sites, but also on the crystal structure, interatomic spacing and crystallite size of the catalytic material. By varying crystal size and phase through molecular engineering, it is possible to manipulate and optimize the catalyst design of titania crystals of controlled size (4–100 nm) and phase which are systematically synthesized by sol–gel hydrolysis–precipitation, followed by hydrothermal treatment [3]. Specifically, 10 nm anatase crystallites due to their greater redox potential present the best photonic efficiency for the photodecomposition of chloroform and trichloroethylene.

Also through supramolecular templating, nanoporous systems can be derived with well-defined pore size and structure, as well as compositional flexibility in the form of particles and thin films. Microporous materials including zeolites and tailored with well-defined pore structures for excellent surface areas and product selectivity are now typically derived through templating with individual molecules. The resulting zeolitic structure which consists of pore opening <1.5 nm allows only small molecules to enter and react, this providing shape and selectivity in separations and catalytic reactions.

Particularly noteworthy is the possibility of creating mesocellular foams produced by templating with triblock copolymers and trimethylbenzene. The resulting materials are composed of uniformly sized, large spherical cells up to 35 nm, which are interconnected by windows to create a continuous three-dimensional pore network. They are attractive for use as catalyst supports in pharmaceutical synthesis as they will permit the diffusion of large substrates through their large open-pore architecture. These porous matrices can also host oxide clusters and active metal and fixate organometallic ligands, offering new possibilities for creating heterogeneous catalyst useful in selective fine chemicals synthesis and asymmetric catalysis [3] and [4].

We could also add that in the field of homogeneous catalysis a supramolecular fine chemistry has been recently established extending the principle of self-organization of the enzyme (catalyst/molecule) to non-biological systems in using supramolecular compounds as catalysts for the shape selectivity of molecules. Such catalysts are formed in situ by self-organization, i.e., chemical bionics [5]. So the latest advances in nanotechnology have generated materials and devices with new physical characteristics and chemical/biochemical functionalities for a wide variety of applications. And chemical engineers and researchers are uniquely positioned to play a pivotal role in this technological revolution with their broad training in chemistry, physical chemistry, processing, systems engineering, and product design.

Increase Selectivity and Productivity by Supplying the Process with a Local "Informed" Flux of Energy or Materials

At a higher microscale level, detailed local temperature and composition control through staged feed and heat supply or removal would result in higher selectivity and productivity than does the conventional approach, which imposes boundary conditions and let a system operate under spontaneous reaction and transfer processes. Finding some means to convey energy at the site (supplying the process with a local "informed" flux of energy) where it may be utilized in an intelligent way is therefore a challenge. Such a focused energy input may be achieved by using ultrasonic transducers, laser beams or electrochemical probes. And to drive the elementary processes within the unit is a challenge but combining microelectronics and elementary processes, e.g. tuning the selectivity by controlling catalytic reactions at the surface of electronic chips is a track being explored.

More Clearly Recognized is the Necessity to Increase Information Transfer in the Reverse Direction, from Process to Man

This means developing all kinds of intelligent sensors, visualization techniques, image analysis and on-line probes giving instantaneous and local information about the process state. This opens the way to a new "smart chemical and process engineering" requiring close computer control, relevant models, and arrays of local sensors and actuators. Field-programmable analog arrays coupled with microreactor technology promise to change the way plants are built, as well as the methods by which their processes are designed and controlled.

Process Intensification: Design of Novel Equipment Based on Scientific Principles and New Operating Modes and Methods of Production

The progress of basic research in chemical engineering has led to a better understanding of elementary phenomena and now makes it possible to imagine new operating modes of equipment or to design novel equipment based on scientific principles.

Process Intensification using Multifunctional Reactors

Such is the case with the "multifunctional" equipment that couple or uncouple elementary processes (transfer–reaction–separation) to increase productivity, selectivity with the desired product or to facilitate the separation of undesired by-products. Indeed in recent years, extractive reaction processes involving single units that combine reaction and separation operations have received considerable attention as they offer major advantages over conventional processes: due to the interaction of reaction and mass and energy transfer, thermodynamic limitations, such as azeotrope, may be overcome and the yield of reactions increased. So the reduction in the number of equipment units leads to reduced investment costs and significant energy recovery or savings. Furthermore improved product selectivity leads to a reduction in raw material consumption and, hence, operating costs. So globally, process intensification through use of multifunctional reactors permits significant reductions in both investment and plant operating costs (10–20% reductions) by optimizing the process. In an era of emaciated profit margins, it allows chemical producers more leverage in competing in the global market place.

There exists a great number of reactive separation processes involving unit operation hybridisation.

The concept of reactive or catalytic distillation has been successfully commercialized, both in petroleum processing, where packed bed catalytic distillation columns are used, and in manufacture of chemicals where reactive distillation is often employed. Catalytic distillation combines reaction and distillation in one vessel using structured catalysts as the enabling element. The combination results in a constant-pressure boiling system, ensuring precise temperature control in the catalyst zone. The heat of reaction directly vaporizes the reaction products for efficient energy utilization. By distilling the products from the reactants in the reactor, catalytic distillation breaks the reaction equilibrium barrier. It eliminates the need for additional fractionation and reaction stages, while increasing conversion and improving product quality. Both investment and operating costs are far lower than with conventional reaction followed by distillation. The use of reactive distillation in the production of fuel ethers such as tert-amyl-methylether (TAME) or methyl-tert-butyl ether (MTBE) or methyl acetate clearly demonstrates some of the benefits. Similar advantages have been realized with the production of high purity isobutene, for aromatics alkylation, for the reduction of benzene in gasoline and in reformate fractions, for the selective production of ethylene glycol which involves a great number of competitive reactions and for selective desulphurization of fluid catalytic cracker gasoline fractions as well as for various selective hydrogenations. The next generation of commercial processes using catalytic distillation technology will be in the manufacture of oxygenates and fuel additives [6].

An alternative reaction–separation unit is the chromatographic reactor, which utilizes differences in adsorptivity of the different components involved rather than differences in their volatility. It is, especially, interesting as an alternative to reactive distillation when the species involved exhibit small volatility differences or are either non-volatile or sensitive to temperature, as in the case, for example, in small fine chemical or pharmaceutical applications. In all these applications, special care has to be devoted towards the choice of the solid phase (sorption selectivity, sorption capacity and catalytic activity). Typical examples for the adsorbents used are activated carbon, zeolites, alumina, ion-exchange resins and immobilized enzymes [7].

Concerning the coupling of reaction and crystallization, there exist myriads of basic chemical, pharmaceuticals, agricultural products, ceramic powders, pigments produced by reactive crystallization based processes (i.e., processes that combine crystallization with extraction to solution mine salts). These separation processes are synthesized by bypassing the thermodynamics barriers imposed on the system by the chemical reactions and the solubilities of the components in the mixture. By combining crystallizers with other unit operations, the stream compositions can be driven to regions within composition space where selective crystallization can occur [8a].

The complementary nature of crystallization and distillation is also explored. Indeed the hybrids provide a route to bypass thermodynamic barriers in composition space that neither the distillation which is blocked by azeotropes and hindered by tangent-pinches in vapor–liquid composition space nor the selective crystallization which is prevented by eutectics and hampered by solid solutions and temperature-insensitive solubility surfaces, can overcome when used separately [8b]. Extractive and adductive crystallization are solvent-based techniques that require distillation columns. They are applied to high melting, close-boiling systems.

Membrane technologies respond efficiently to the requirement of so-called process intensification. Because they allow improvements in manufacturing and processing, substantially decreasing the equipment-size/production-capacity ratio, energy consumption, and/or waste production and resulting in cheaper, sustainable technical solutions. The paper by Drioli and Romano [9] documents the state of the art well involving progress and perspectives on integrated membrane operations for sustainable industrial growth. This technology can respond to the strongly increasing demand for food additives, feeds, flavors, fragrances, pharmaceuticals, agrochemicals, etc. Phase-transfer catalysis can also be realized in membrane reactor configurations, immobilizing the appropriate catalysts in the microporous structure of the hydrophobic membrane. Catalytic membrane reactors are also proposed to selective product removal to remove equilibrium limitations, i.e., catalytic permselective or non-permselective membrane reactors, packed bed (catalytic) permselective membrane reactors, fluidized bed (catalytic) permselective membrane reactors. For more general applications material scientists must solve the problem of providing inorganic membranes of perfect integrity involving mechanical and thermal stability and which will allow large fluxes of desired species and second, chemical engineers must figure out the heat transfer problem which now threatens successful scale-up.

Finally, though multifunctional reactors are not quite new to the process industries, i.e., absorption or extraction with chemical reaction, only recently reactors incorporating several "functions" in one reactor have been formally classified as being multifunctional and the large benefits obtained in integrating progress of knowledge at different scale and time-lengths have been acknowledged by the process industries. This was illustrated by the first international symposium on multifunctional reaction in 1999 [10]. But to achieve optimal performance with multifunctional reactors, it is important to lead a scientific approach to understand where the integration of functionalities does occurs, as explained in Fig. 5[11] in the case of a catalyst particle in a reacting medium.

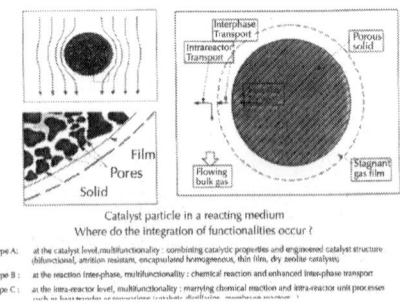

Catalyst particle in a reacting medium
Where do the integration of functionalities occur ?

Type A: at the catalyst level,multifunctionality : combining catalytic properties and engineered catalyst structure
           (bifunctional, attrition resistant, encapsulated homogeneous, thin film, dry zeolite catalysts)

Type B :   at the reaction inter-phase, multifunctionality : chemical reaction and enhanced inter-phase transport

Type C :   at the intra-reactor level, multifunctionality : marrying chemical reaction and intra-reactor unit processes
           such as heat transfer or separations (catalytic distillation, membrane reactors)

Figure 5: Process intensification using multifunctional reactors [11].

However, we will mention more generally that the use of hybrid technologies encountered in a great number of multifunctional reactors is limited by the resulting problems concerning control and simulation, i.e., the interaction between simultaneous reaction and distillation introduces more complex behaviour involving the existence of multiple steady states and output multiplicities corresponding to different conversion and product selectivity, compared to conventional reactors and ordinary distillation columns. This leads to interesting challenging problems in dynamic modelling, design, operation, and strong non-linear control.

Process Intensification using New Operating Modes

The intensification of processes may be obtained by new modes of production which are also based on scientific principles. Indeed new operating modes are in the laboratory and/or pilot stage: reversed flow for reaction-regeneration energy efficient coupling of endo- and exothermic reactions, countercurrent flow and induced pulsing flow in trickle beds, unsteady operations, cyclic processes, extreme conditions, pultrusion, low-frequency vibrations to improve gas–liquid contacting in bubble-columns, high temperature and high-pressure technologies, and supercritical media, and use of composite structured packings achieving low pressure drop through vertical stacking of catalyst, are now seriously considered for practical application.

Process Intensification using Microengineering and Microtechnology

Current production modes also are and will be more and more challenged by decentralization, modularization and miniaturization. Microtechnologies recently developed, especially in Germany (i.e., IMM, Mains and Forschnungszentrum, Karlsruhe) and in USA (i.e., MIT and DuPont) lead to microreactors, micromixers, microseparators, micro-heat-exchangers, and microanalyzers, making possible accurate control of reaction conditions with respect to mixing, quenching, and temperature profile.

Miniaturization of chemical analytic devices in micro-total-analysis-system (μTAS) represents a natural extension of micro-fabrication technology to biology and chemistry with clear applications in combinatorial chemistry, high-throughput screening, and portable analytical measurement devices. Also the merging of μTAS techniques with microreaction technology promises to yield a wide range of novel devices for reaction kinetic and micromechanism studies, and on-line monitoring of production systems [12].

Microreaction technology is expected to have a number of advantages for chemical production [13] and [14] as the high heat and mass transfer rates possible in microfluidic systems allow reactions to be performed under more aggressive conditions with higher yields that can be achieved with conventional reactors. Also new reaction pathways considered too difficult in conventional microscopic equipment, e.g., direct fluorination of aromatic compounds, could be pursued because if the microreactor fails, the small amount of chemicals released accidentally could be easily contained. And the presence of integrated sensor and control units could allow the failed microreactor to be isolated and replaced while other parallel units continued production. Also these inherent safety characteristics could allow a production scale systems of multiple microreactors enabling a distributed point-of-use synthesis of chemicals with storage and shipping limitations, such as highly reactive and toxic intermediates (cyanides, peroxides, azides) [13].

Moreover, scale-up to production by replication of microreactors units used in the laboratory would eliminate costly redesign and pilot plant experiments, thereby shortening the development time from laboratory to commercial-scale production. This approach would be particularly advantageous for pharmaceutical and fine chemicals industries where production amounts are often less than a few metric tons per year.

Also it was proposed a new concept for high-throughput screening (HTS) experiments for rapid catalyst screening based on dynamic sequential operations with a combination of pulse injections and micromachined elements [15]. The authors describe a new concept to achieve HTS of polyphasic fluid reactions for two test reactions, a liquid–liquid isomerization of allylic alcohols and a gas–liquid asymmetric hydrogenation. The principle used for the test microreactor is a combination of pulse injections of the catalyst and the substrate, an IMM static micromixer with negligible volume and residence time less than 10^{-2} s, and a tubular reactor. The two scanning electron microscopy images show the micromixer, in which 2 × 15 interdigitated microchannels (25 µm width) with corrugated walls are clear (Fig. 6). The pulses mix perfectly in the micromixer and the liquids or the gas–liquid mixtures thereby emulsify and behave as a reacting segment, which then travels along the tubular microreactor.

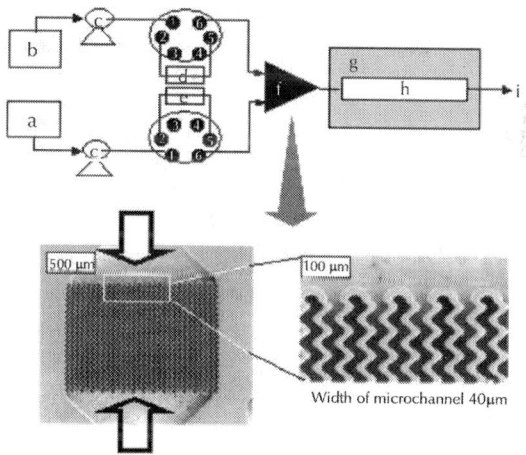

Experimental set up : a water reservoir, b.n-heptane reservoir , c.high pressure liquid pumps, d and e, HPLC type injection valve equipped with 200 ml and 1 ml loop, f.MICROMIXER. G thermoregulated bath, h. tubular stainless steel reactor (80cm,length, 0.4cm i.d) i outlet analytics, J SEM image of the mixing micro element showing the 2x 15 interdigited microchannels (25?m width corrugated walls

Figure 6: The IMM micromixer for high-throughput screening [15].

Collection at the outlet of the reactor and analysis afford the conversion and selective data. The catalyst library was then screened. The results led to the selection of the best catalyst showing activity towards a large class of allylic alcohols. Similar results which were obtained in a microreactor and in traditional well mixed batch reactor (40 cm³) prove the validity of the concept (Fig. 7).

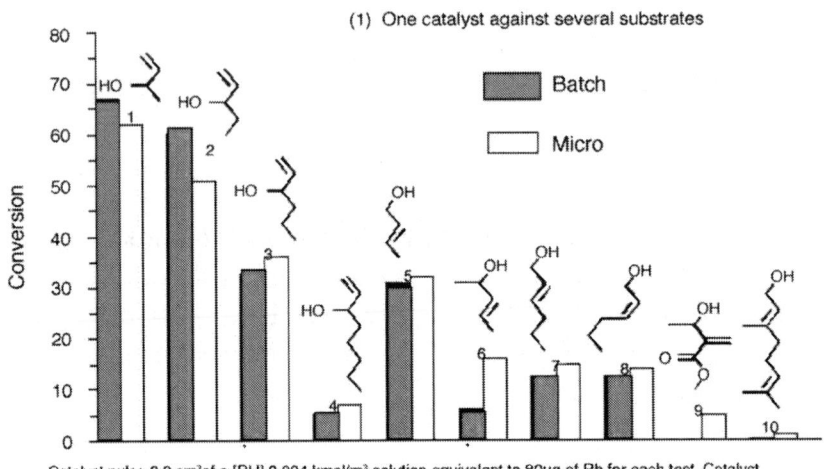

Catalyst pulse 0.2 cm³of a [RH] 0.004 kmol/m³ solution equivalent to 80μg of Rh for each test. Catalyst RhCl³/TPPTS/NaOH. Rh/TPPTS = 5. Selectivity . 95%.70°C. Flow rate : Aqueous phase 5 cm³/min Organic phase 1 cm³/min [substrate] 0.1 km ol/m³ Residence time 100 s.

Figure 7: Comparison with traditional equipment (batch reactor) [15].

In term of catalyst and time consumption per test, the numerous tests for the liquid–liquid isomerization were performed twice, to test for reproducibility, using only 1 or 2 μmol of metal and over a total screening time of 1 h. The test for the gas–liquid asymmetric hydrogenation showed similar features (down to 0.2 μmol of catalyst, and 3–5 min per test). Throughput testing frequencies of more than 500 per day are achievable, albeit with computer control of the apparatus. Using these microreactors for dynamic, high-throughput screening of fluid–liquid molecular catalysis offer considerable advantages over traditional parallel batch operations: ensuring good mass and heat transport in a small volume, reduced sample amounts (to μg levels), a larger range of operating conditions and simpler electro-mechanical moving part.

Product Design and Engineering: Manufacturing End-use Properties: Development of Multidisciplinary Product-oriented Engineering with a Special Emphasis on Complex Fluids and Solids Technology

This is the answer for the nowadays ever-growing market place demand for sophisticated products combining several functions and properties: cosmetics, detergents, surfactants, bitumen, adhesives, lubricants, textiles, inks, paints, paper, rubber, plastic composites, pharmaceuticals, drugs, foods, agrochemicals, and more.

This product design and engineering (synthesis of properties), is the translation of molecular structure into macroscopic phenomenological laws in terms of state variables and in practice it mostly concerns complex media and particulate solids. Indeed complex media such as non-Newtonian liquids, gels, foams, hydrosoluble polymers, colloids, dispersions, emulsions, microemulsions, suspensions for which rheology and interfacial phenomena play a major role are often involved. Also involved are the so-called "soft solids", systems which have a detectable yield stress, such as ceramic pastes, foods or drilling muds.

Product design concerns also particulate solids encountered in 70% of the process industries. This involves the creation and the control of the particle size distribution in operations such as crystallization, precipitation, prilling, generation of aerosols and nanoparticles as well as the control of the particle morphology and the final shaping and presentation in operations such as agglomeration, calcination, compaction, and encapsulation. Both types of operations need a better understanding as they control the end-use property and quality features, such as taste, feel, smell, colour, handling properties, sinterability or biocompatibility. Product design and engineering concerns also solids considered as vehicles of condensed matter from the perspective of solventless processes or non-passive "intelligent solid" to accomplish intelligent functions such as controlled reactivity or programmed release of active components that may be obtained by multiple layer coatings.

The quality and properties of emulsified or past like and solid products is determined at the micro- and nano-level. Therefore to be able to design and control the product quality and make the leap from the nano-level to the process level, chemical and process engineering involved with structured material have to face many challenges in fundamentals (structure–activity relationships on molecular level, interfacial phenomena, i.e., adhesive forces, molecular modelling, i.e., equilibria, kinetics, product characterization techniques, etc.), in product design (nucleation growth, internal structure, stabilization, additives, etc.), in process integration (simulation and design tools based on population balance) and in process control (sensors and dynamic models).

For illustration, we may cite the control of the quality of microemulsions for foodstuffs containing microorganisms that could spoil and whose growth can be prevented by enclosing them in a water-in-oil emulsion of aqueous droplet size not significantly larger than 1 µm and of a narrow size distribution, which namely characterizes the product quality [16]. Such miniemulsions can only be generated in high-pressure homogenizers with a high-energy input and customized nozzle geometry. However, the droplets generated must not coalesce during emulsification which makes it necessary to find emulsifier systems which also stabilize the droplets sufficiently fast. So, in modelling the emulsification process, the overall process has to be divided into two substeps: generation of droplet by mechanical energy and stabilization of the droplets before they re-coalesce. The resulting product quality is determined not only by how well the dispersed phase has been broken up into small droplets but also by how well the equipment, process conditions and emulsifier have been matched to one another. Thus, the kinetics of the molecular process determines whether the desired end-product properties will really be achieved, even if the required droplet size had been achieved in the first substep.

Complementary in topics such as microemulsions for chemical, food and pharmaceutical industries (drug delivery systems), it should be emphasized recent investigations on monodisperse emulsion formation with micro-fabricated micro channel (MC) array, called straight through microchannel, i.e., silicon array of elongated through-holes for monodisperse emulsion droplets [17]. Such oblong straight through MC equipment allows getting monodisperse oil in water emulsion droplets with average diameter of 32.5 μm and a coefficient of variation of 1.5% verifying their excellent monodispersity (Fig. 8). Such monosized droplets in emulsions have advantages for control of their physical and functional end-use properties, stability and application to other processings.

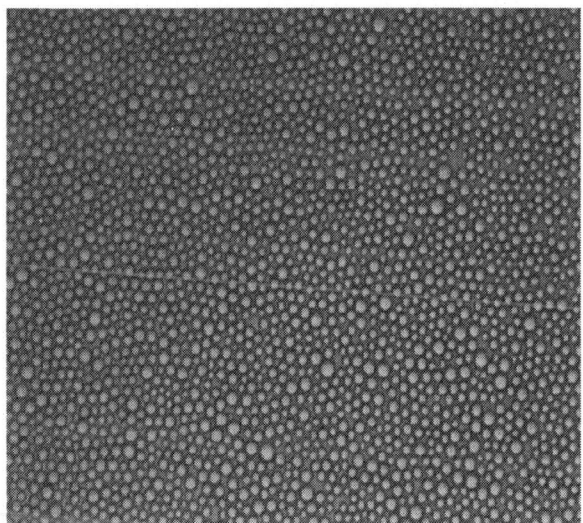

Figure 8: Monodispersed emulsion formed with micro-fabricated micro-channel array.

For topical delivery especially on the skin, novel multiple lipidic systems account for sustained release and optimized stabilization of active ingredients as well as drugs. Topical delivery for cosmetic products combine aspects of optimized skin release of actives and an optimized match to sensorial features of a product. Prominent examples for preparation of such kind are multiple emulsions of the water-in-oil in water type (W/O/W type), produced by the partial-phase soluinversion technology (PPSIT), and solid lipid nanoparticles (SLN, lipopearls) and multicompartment solid lipid nanoparticles (MSLN).

Multiple emulsions based on the PPSIT technology (Fig. 9) combine protecting and occluding effects of classical W/O emulsions and easy application feature of classical O/W formulations. Besides, the W/O/W base as such already shows excellent skin caring properties, as exemplified by improving skin's microrelief, short-, mid-, and long-term moisture holding capacity (adaptogenic moisturization) and skin firmness improvement. Such multiple emulsions are manufactured by novel one step manufacturing technology even facilitating industrial scaling up to large scale (up to 1 t batches).

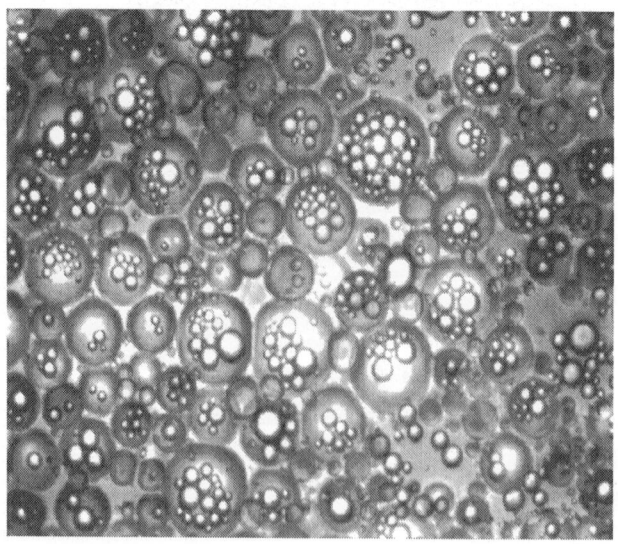

Figure 9: Multiple W/O/W emulsion manufactured by partial-phase solu-inversion technology [18].

Formulation of oxidation-instable ingredients such as lipoic acid and retinol are preferentially stabilized in solid lipid nanoparticles

(SLN) suspensions, which can either coat the instable materials as solid shell or even can entrap additional solving oil compartments to be detected active (Fig. 10). SLN particles can be manufactured based on proloxamer derivatives as well as non-ethoxylated lipids, such as compritol or dynasan. High-pressure homogenization reveals also ultra-narrow particle size distribution in the nanometer range and an excellent stabilization of lipophilic ingredients such as Ubiquinone Q10 and Vitamin E and derivatives. Due to the solid character of this carrier, active ingredients can either be protected against oxidation and hydrolysis.

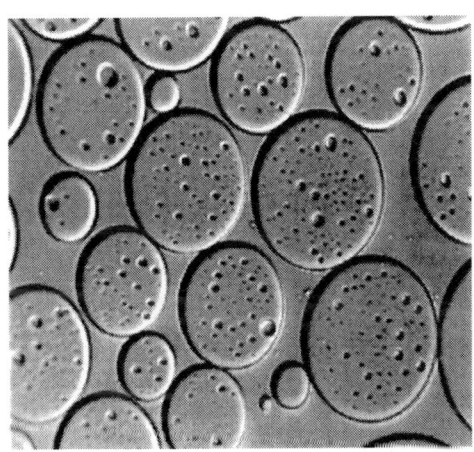

Figure 10: Solid lipid nanoparticles.

For other illustration we may cite the control of the shape and size of crystals in an industrial crystallization process. It has been shown that much improved process control, both in terms of crystal purity and a defined size distribution could result by detailed computer studies of the crystallization processus which can be remarkably changed by the presence of small traces of foreign substances such as unwanted by-products in the feed solution [19] and [20].

Indeed in order to understand the mechanisms causing these changes in crystals size and shape so that it is possible to utilize them in a controlled manner, one must explain the structure–activity relationships on a molecular level. And with computer simulations, diagrams of the molecular structure of the most important crystal surfaces can be generated from X-ray crystal structure data.

Also in the same computer simulations, contaminant molecules or molecules with an expected beneficial effect on crystallization processus can be placed on each crystal surface and their adsorption energy calculated. If the hypothesis is that the growth rate of surface decreases with increasing adsorption energy, and by comparing relative adsorption energies the modified crystal shape to be expected can be predicted. This was illustrated with the results of crystallization from a feed ammonium sulphate solution containing dye amaranth. It was shown that the molecule amaranth is adsorbed onto 0 0 1 surface of ammonium sulphate crystals with the highest adsorption energy in comparison with the other crystal surface. And according to the calculations, the somewhat block-shaped crystal produced in the pure system becomes a flat shaped crystal having a large 0 0 1 surface area, which was experimentally verified (Fig. 11[19]). Comparable prediction was obtained in the case of a feed ammonium sulphate solutions containing 50 ppm Al^{3+} [20].

Modification of Crystal Shape
by Additives

Pure System   70 mm of dye amarnath

Block-shaped crystal in pure ammonium sulphate solutions and flat shaped crystal having a large 001 surface area ni a solution containing dye amarnath

Figure 11: Product-oriented engineering: controlled crystallization process [19].

So many quality features can only be designed in a targeted way if the molecular processus are understood at this level. And as shown by this example the analysis—both by theoretical and by experimental means—must be carried out down to the molecular level to obtain results of real value for understanding the relationship between a certain set of product qualities and the physical product state: indeed it is well known that two pain killing tablets may have the same chemical composition, but different routes of production may lead to different crystallinity and porosity profiles and therefore to totally different dissolution and solubility properties, namely different bioavailabilities.

New approaches offer now the possibility of accurate predicting the effect of solvents or impurities on crystal shape [21] and recent models recognize the significance of interfacial phenomena in crystal shape modelling, and lead the way for future developments, such as new simulation and/or group contribution methods for interfacial free energy production.

Another illustration of this multilevel research effort in crystallization has been proposed in the area of electrical engineering: microelectronics. It concerns correlations between operating conditions and microstructure of low pressure chemical vapor deposit (LPCVD) silicon based films prepared from silane, SiH_4. The main aim is the development of a rigorous simulation model for the interpretation of the layer growth experimental data by taking into account both the mass transfer resistance at boundary layer and the solid layer growth kinetic expression [22]. The four blocks of accessible knowledge for a LPCVD process are presented in Fig. 12. Specialists in material sciences may relate microstructure and properties of thin films while specialists in chemical engineering can correlate and simulate the macroscopic operating conditions of a CVD reactor and the local conditions inside the reactor. It is clear that any relationship established between blocks 2 and 3 affords an intrinsic property of the system analyzed which is independent of the equipment used and which allows to treat the complete design or optimization problems completely by theoretical methods. For example it should be possible to treat, first, the kinetic questions such as to find sets of conditions producing the desired thickness with the desired degree of uniformity. In a second step, one could solve the questions of material structure (amorphous, partially or totally crystallized silicon) end-use properties.

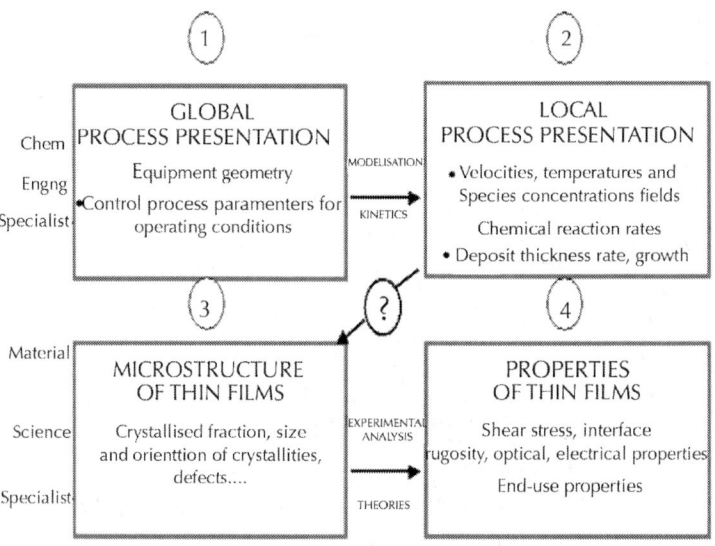

Figure 12: The four blocks of an accessible knowledge for a CVD process: the aim, relation between 2 and 3 intrinsic property of the system (independent of the reactor geometry) [22].

The thin CVD layers were produced in the so-called hot wall tubular reactor (2 m length, 2.2 m in diameter). Numerous characterization methods have been used by the authors to study the layers microstructural evolutions in function of their elaboration conditions. The local state of the reactor was simulated by a model CVD including the gaseous flow hydrodynamic and the mass transfer and chemical reaction parameters both in heterogeneous phase (gas–solid interface) and in homogeneous phase (solid phase). Mechanism existing at surface substrate and nucleation and crystal growth phenomena together with the dynamic characteristic of the microstructure formation were modellized with simple geometric and statistic approach based on concepts of the mechanics of continuous media (instead of using molecular dynamic models requiring too much computer power). The final model that necessitates the local conditions of elaborations at surface substrate and nucleation and crystal growth laws, is able to calculate the thickness, the fraction of crystallized silicon and the space distribution of silicon crystallites in the deposit layer. So with such a model it is possible to predict the microstructure of LPCVD films from operating conditions such as gas phase components, temperature, pressure and initial substrate nature or surface state. A comparison between the transmission electronic microscopy micrographies of the experimental silicon films from pure silane for different temperatures and deposit durations and the simulation of the model has shown a good approximation for the crystallized fraction x_c and the space distribution of crystallites (Fig. 13). The results also emphasize the dynamic characteristics of film microstructure. Similar results were obtained for complex SIPOS SiOx deposit elaborated from silane and nitrogen protoxide and in situ boron silicon deposits from silane and boron trichloride [23]. It is interesting to note that the quality of such simulation results demonstrates the validity of the approach proposed and suggests the way now opened to develop a complete product design and engineering or engineering of materials elaboration, able to predict the kinetic and structural characteristics of LPCVD films by numerical simulation.

Finally, much progress has been realized these last few years for the product design and the control of the process using the scientific methods of chemical engineering. Thermodynamic equilibrium states are examined, transport processes and kinetics are analyzed separately and these are linked again by means of models with or without the help of molecular simulation and finally with the help of computer tools for simulation, modellization and extrapolation at different scales for the whole supply chain (BASF, Unilever, Degussa, Astra Zeneca, Nestlé, etc.).

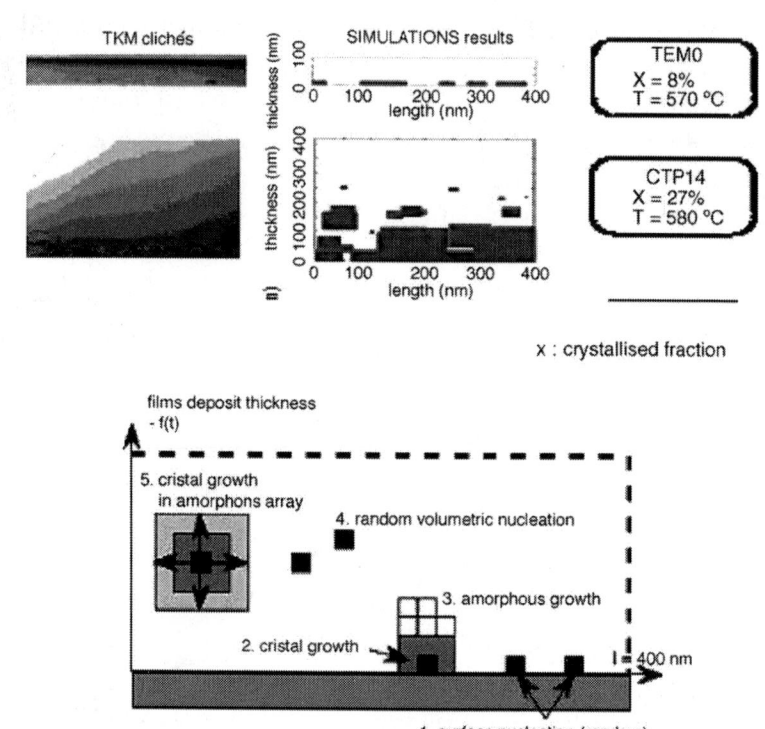

Figure 13: Microstructure of silicon LPCVD films from silane [22].

But how can operations are scaled up from laboratory to plant? Will the same product be obtained and will its properties be preserved? What is the role of equipment design in determining product properties? How can the optimal interactions between product and process design be explored? How can we validate and test the desired functional properties (i.e., controlled drug release, enhanced bituminous or textile behaviour, skin improvements in using cosmetic creams, etc.) of the product in use? Indeed as underlined many times, the control of end-use properties is a key issue for which general scale-up rules are still lacking. This requires for chemical engineering specialists and their systemic approach a close cooperation with specialists in physical chemistry, biology, mechanics and mathematics to develop this new "systemic" physical chemistry and biology where qualitative explanation will be translated with the help of fine modellization into formal laws for process development. This leads to the fourth main objective.

Implement Multiscale and Multidisciplinary Computational Chemical Engineering Modelling and Simulation to Real-life Situations: from the Molecule to the Overall Complex Production Scale into the Entire Production Site

We have emphasized the necessary multidisciplinary and multiscale integrated approach applied to the triplet processus–product–process to scale from the nano- and microscales of end-use properties of the product to the mesoscale of the equipment manufacturing the product.

Computers have opened the way for chemical and process engineering in the modelling of molecular and physical properties on the nano- and microscopic scales. For the molecular modelling, application of the principles of statistical molecular mechanics computational techniques (Monte Carlo and molecular dynamics) and quantum mechanics constitute an area for the problem-oriented approach of chemical and process engineering. Indeed molecular modelling starts from a consideration of microscopic structure and molecular interactions in a material system and derives thermodynamic, transport, rheological, mechanical, electrical, electronic or other properties through rigorous deductive reasoning bases on the principles of quantum and statistical mechanics. Compared to more phenomenological approaches (e.g. correlations of the group contribution type) it offers the advantages of greater, generality and reliability.

It is clearly impossible to cover all directions of present-day molecular modelling researches involved in a wide spectrum of problems in the chemical and material sciences. Vapor–liquid equilibria, vapor–liquid–liquid equilibria, liquid–solid equilibria, supercritical solution properties; amphiphiles; polymers at interfaces; adsorption on surfaces and influence of impurities; microporous materials or ceramics structures; transport properties such as viscosity, diffusivity, thermal conductivity can be calculated today by molecular modelling based on information from thermodynamic, kinetic and rheological data banks.

There is no doubt that molecular modelling is now playing an increasingly important role in future chemical and process engineering research and practice [24]. Recent advances in the fundamental molecular sciences and in computer hardware and numerical algorithms have greatly accelerated its development (see Fig. 14 and Fig. 15, where it is shown that the number of transistors on a sliver of silicon would double every 2 years).

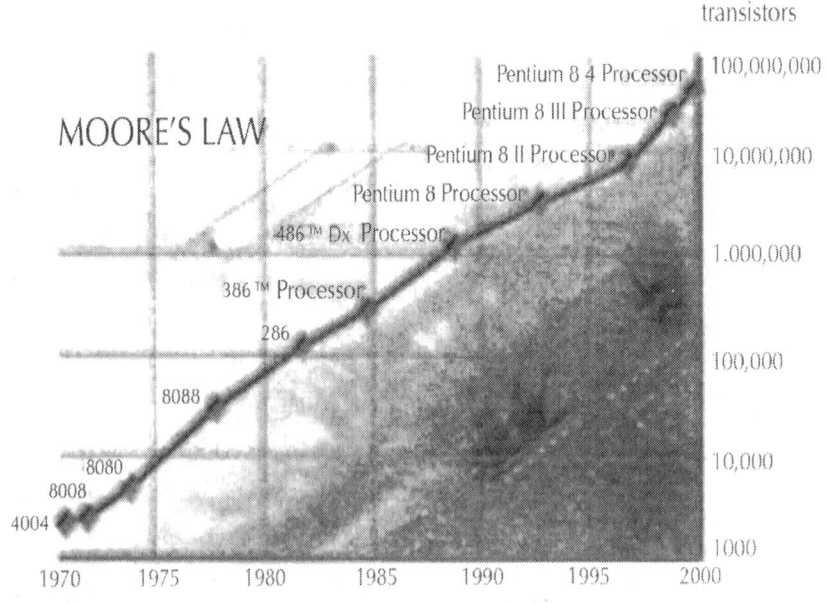

Figure 14: Moore's law number of transistors on silicon would double every 2 years.

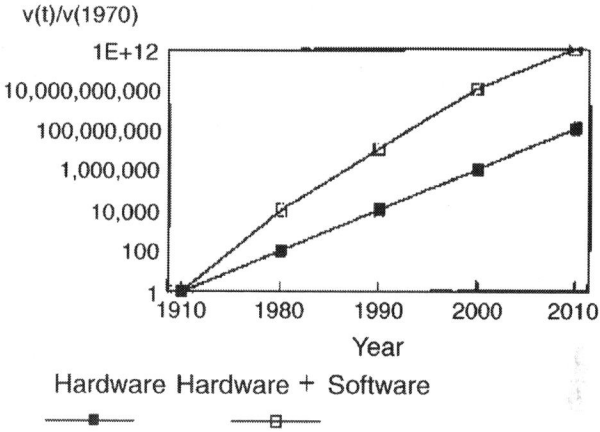

v(t)/v(1970)

• Speed of Electronics computation (H+S development) roughly doubled every year since 1970 and is expected to continue in the coming decade

• Idem about hard-disk drive capacities :
        0.2 gigabyte /cm^2 (1999) 15 gigabytes /cm^2 (2010)
        150 gigabytes /cm^2 (after 2010) with holographic memory technology
        substitute magneto-resistive technology

Figure 15: Computing speed acceleration.

And through the interplay of molecular theory, simulation, and experimental measurements evolves a better quantitative understanding of structure–property relations, which, when coupled to macroscopic chemical engineering science, can form a basis for new materials and process design.

Furthermore, turning to the macroscopic scale, dynamic process modelling and process synthesis are being also increasingly developed. Indeed one must remember the targeted products in question are generally not mass-produced products but ones which are produced in small batches and just in time for delivery to the customer whose needs are constantly changing and evolving. And to be competitive under these conditions, it is particularly important to analyze and optimize the supply chains for which we are interested in the time that individual process steps take, and these have to be simulated and evaluated also in terms of costs.

25.   M. Pons, B. Braunschweig, K. Irons, J. Koller, A. Kuckelberg, P. ¨ Roux, CAPE-OPEN (CO) standards: implementation and maintenance, in: Proceedings of the 3rd European Congress of Chemical Engineering, Computational Engineering Session 7, Poster 188, Nurnberg, 26–28 June 2001. ¨

26.   R. Gani, Chemical product design: challenges and opportunities, Comput. Chem. Eng. 26 (2002) 984.

27.   D. Frenkel, B. Smit, Understanding Molecular Simulation, Academic Press, San Diego, 2002.

28.   L.F. Gladden, Magnetic resonance: ongoing and future role in chemical engineering research, AIChE J. 499 (1) (2003) 2–9.

29.   N.M.R. Koptyug IV, Imaging: a powerful tool kit for catalytic research, in: Proceedings of the XVI International Conference on Chemical Reactors, Berlin, 1–5 December, 2003, pp. 15–24 (abstract No. PL14).

30.   C. Boyer, A.M. Duquenne, G. Wild, Measuring techniques in gas–liquid and gas–liquid–solid reactors, Chem. Eng. Sci. 57 (2002) 3185–3215.

Advanced Chemical Processing using Microspace

Kazuhiro Mae

Department of Chemical Engineering, Kyoto University, Katsura, Nishikyo-ku, Kyoto 615-8510, Japan

INTRODUCTION

Recently, the merits of microreaction technology have been recognized on a worldwide level. Active microreactor development started in the 1990s, mostly to obtain useful tools for chemical analysis. This led to the development of several useful lab-on-chips and μ-TAS (Harrison et al., 1993, Reyes et al., 2002 and Auroux

MICRODEVICES FOR CHEMICAL PRODUCTION

To utilize the advantages of microtechnology in actual processing, precisely designed microstructured devices are required. Various microdevices have been developed by now (Hessel et al., 2004a). Micromixing is thought to be the key operation for efficient microprocessing (Ehrfeld et al., 1999). Microdevices utilizing micromixing can be categorized into three types according to the driving force of mixing: molecular diffusion driven, momentum driven, and external force (e.g. electric force) driven devices.

In microreactors, mixing induced by molecular diffusion becomes dominant, since reactor miniaturization leads to low Reynolds numbers in reactor channels. Therefore the diffusion length between reactants must be reduced to achieve efficient mixing. One way to achieve fast mixing is to feed small fluid segments made by splitting reactant fluids using unique microchannel geometries into the mixing chamber. Examples of micromixers using this mixing method are the interdigital mixer (Ehrfeld et al., 1999), and the multi-stream mixer with focusing after confluence (Löb et al., 2004 and Wang et al., 2005). When only this principle is used for mixing, shortening the diffusion length by channel size reduction is essential to achieve fast mixing. However, such channel size reduction also leads to a high pressure drop in the channel and thus to a limited flow rate. This results in a low productivity and various difficulties in operation. Therefore, another principle to enhance mixing performance is needed, especially when industrial production where high throughput is required.

To overcome low productivity, another type of micromixer was developed which uses momentum as the driving force for mixing. In laminar flow mixing, applying high shear rates to reactant fluids is important to shorten mixing time (Mohr et al., 1957). Colliding two fluid streams is the simplest method to utilize this mixing principle. T- and Y-shape microchannels are typical examples (Engler et al., 2004, Gobby et al., 2001, Ookawara et al., 2004 and Wong et al., 2004). Combining the two principles mentioned above is an effective method to split reactant fluids into small fluid segments. This method enables us to avoid the usage of extremely small channels to produce small fluid segments, so high pressure drops do not occur. Micromixers based on this combined mixing principle have been developed (Hessel et al., 2005a and Yang et al., 2004). In the mixer developed by Nagasawa et al. (2005) and Nagasawa and Mae (2006), reactant fluids are first divided into fairly small fluid segments using relatively small channels, and next the fluid segments collide at a single point and shear is applied to them. After collision, the fluid segments are recombined, and the reactant fluids enter the channel of the outlet plate. This is only one example, and various types of micromixers designed using this combined mixing principle have been successfully presented.

Based on such micromixers, various microreactors have been developed. For liquid reactions the simple combination of a micromixer and a microtube is quite popular. Fast mixing, fast heat exchange and the strict controlling of residence time can be achieved using this simple system. Another popular type is a module type microreactor. Various stainless steel plate modules, each having carefully designed microchannels engraved within them, were stacked according to the reaction scheme the microreactor is going to be used for (John et al., 2003). Each module is used for a certain purpose, for example mixing, heat exchange, residence time controlling, quenching or separation. By this system we can set up a suitable reactor to accomplish a certain reaction, by simply combining modules.

For particle production, several special microreactors have been developed. The first type is a jet mixer reactor, in which jet flows of reactants from microchannels are mixed in a free space (Hessel et al., 2003). The next type is a multi-layer annular flow reactor (Nagasawa and Mae, 2006). The outer annular flow acts as a sheath fluid which prevents the adhesion of particles to the reactor wall. The details of this type of reactor will be described later. The third type is a segmented flow reactor (Song et al., 2003, Zheng et al., 2003, Zheng et al., 2004, Burns and Ramshaw, 2001 and Henkel et al., 2004). In this type of microreactor, a slug flow is formed by droplets. The reaction occurs only within the droplet, so the residence time can be controlled by adjusting the flow rate of the slug flow. The adhesion of particles to the reactor wall can be avoided, as the particles are completely confined within the droplets. Since the mixing rate of reactants within the droplets is extremely fast, because of rapid circulation caused by friction between the wall and the droplets, such segmented flow operation is very attractive (Tanthapanichakoon et al., 2006).

Catalytic microreactor for gas phase reaction can be classified into two types: one is packed type and the other is stacked type with wall catalyst. In a microreactor using a wall catalyst, catalyst exchange becomes a serious problem because of its high cost. To overcome this issue, assembly type reactors have been developed (Wolfrath et al., 2003; Maki et al., 2005). In this type of reactor, catalyst elements are simply inserted into the reactor, and the microspace formed between the catalysis element and reactor wall is utilized as the reaction field. A large interfacial area per unit volume can be achieved which is attractive for two phase reactions. A falling film type and a microbubble column type reactor were presented by IMM (Hessel et al., 2000 and Haverkamp et al., 2001). Interfacial area per unit volume values as high as $20,000 \ m^2 \ m^{-3}$ can be obtained, and a significant increase in the reaction rate of gas–liquid reaction was observed. Thus, various types of micromixers and microreactors have been developed in the past decade. In summary, a microdevice is a functional device to perform a specific operation intensively. This means that the performance of the device strongly depends on its design. In this sense, a microdevice should be designed based on the logic of the reaction scheme it is going to be used for. In the following sections, the logic for designing and operating microdevices, which are sure to lead to the establishment of a new production technology, were surveyed.

AS A NEW PRODUCTION TECHNOLOGY IN ORGANIC SYNTHESIS—PRECISE CONTROLLING OF HOMOGENEOUS REACTIONS UNDER SEVERE CONDITIONS

Unique features which can be achieved by utilizing microspaces, such as fast mixing, rapid heat exchange, etc. can be used for the controlling of difficult reactions in organic synthesis. Reactor miniaturization permits us to conduct reactions under more precisely controlled conditions, and is thought to bring along the following merits (Hessel et al., 2004a): (1) shortening of reaction time; (2) solvent-free operation; (3) new synthesis routes; (4) controlling of unstable intermediates; (5) safe operation under explosive and thermal runaway conditions; (6) omission of catalysts and reagents. Typical examples are summarized inTable 2. More examples can be found in the review by Hessel et al. (2004a).

Table 2: Examples of organic syntheses using microreactors

Category	Kind of reaction	Comments	Ref.
(1) Shortening reaction time	Many reports	See review article	Hessel et al. (2004a)
(2) Solvent free	Bromination of thiophenol	Room temp., yield 86% (Batch: 50%)	Löb et al. (2004)
Neat reaction	Production of Bis-phenols	Phenol/formaldehyde: stoichiometric ratio	Daito et al. (2006)
(3) New route	Direct fluorination from elemental F	Residence time: 100 ms, safely operation	Chambers and Spink (1999)

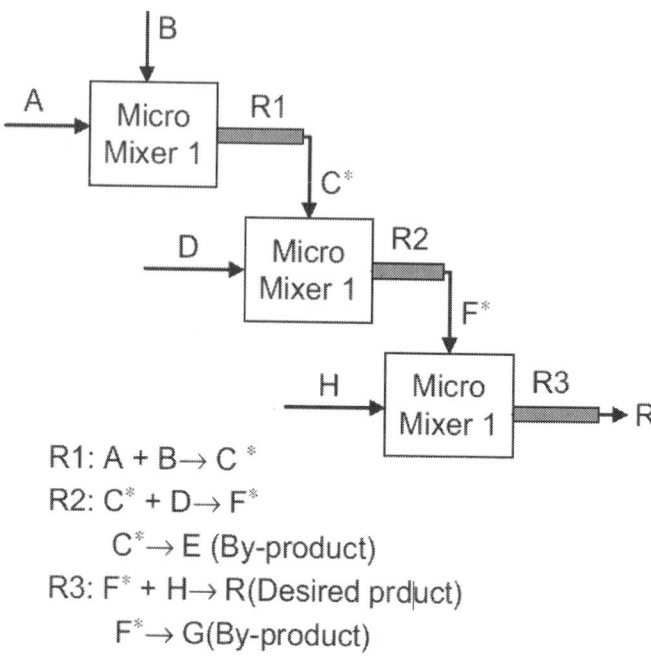

R1: A + B→ C *
R2: C* + D→ F*
       C*→ E (By-product)
R3: F* + H→ R(Desired prduct)
       F*→ G(By-product)

Figure 1: Design of reactor configuration followed by reaction mechanism.

As another distinct feature, properly designed microreactors can be operated easily and safely, and even explosive reactions can be conducted under severe conditions. As an example, the selective oxidation of 2-methylnaphthalene with 60% of peracetic acid at 110 °C was performed under 0.6 MPa in the presence of a Pd catalyst (Yube and Mae, 2005). Usually, the reaction is conducted at a low temperature where the unfavorable reaction with OH radicals becomes dominant as shown in Fig. 2, since peracetic acid is highly explosive. Microreaction technology allows us to operate safely under high temperature and high pressure, so oxidation can be conducted under favorable reaction conditions. As shown in Fig. 2, VK3 was successfully obtained at yields up to 60% by operating at high temperatures. Thus, microreactors provide us a different reaction field in which the favorable reaction with OH$^+$ ion can be accelerated. This indicates that microreaction technology allows us to use new reaction paths in organic synthesis.

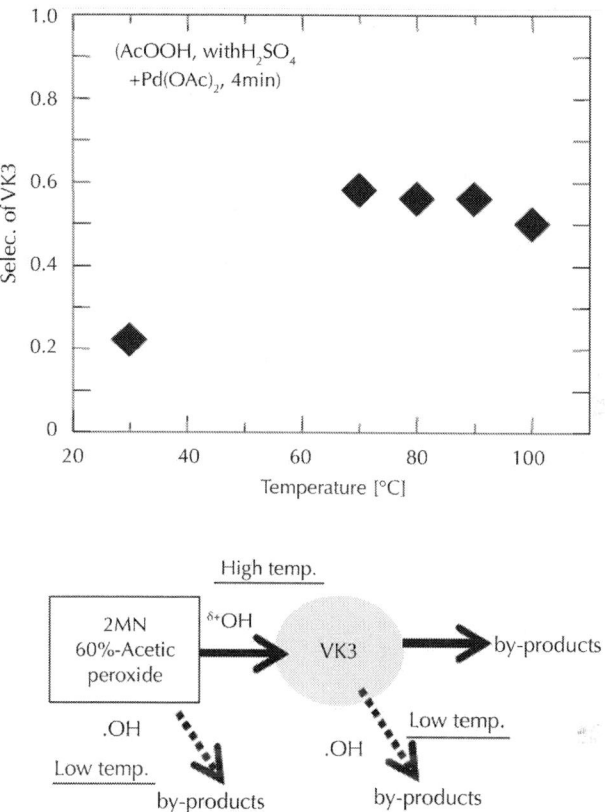

Figure 2: Oxidation of 2-methylnaphthalene with acetic peroxide.

The high heat exchange ability of microdevices can be used to control multiple reactions. If the Arrhenius plots of each reaction can be obtained, then reactor configuration and operation conditions for increasing the target product can be determined. For example, let us consider a parallel reaction represented by the Arrhenius plots shown in Fig. 3. From the plots, we find that the formation of the desired product, R, is favorable at higher temperatures. From this information, we can construct a reactor configuration to achieve rapid heating using microdevices. Thus, microreaction technology is a build-up technology based on functional microdevices for the strict controlling of chemical and physical conditions required from the reaction mechanism.

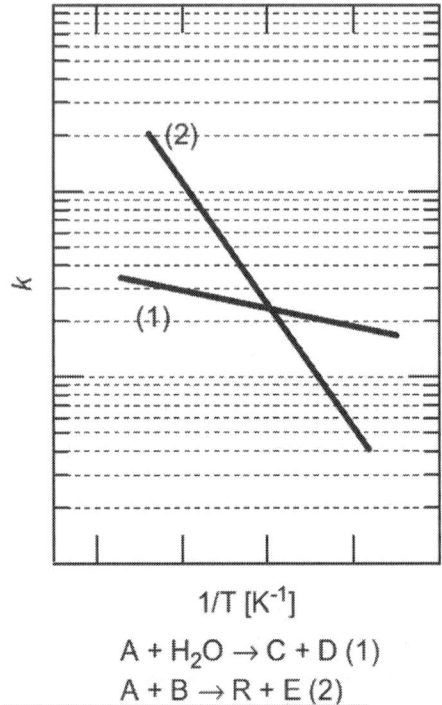

$$A + H_2O \rightarrow C + D \ (1)$$
$$A + B \rightarrow R + E \ (2)$$

Figure 3: Example of Arrhenius plots for parallel reaction.

AS A POWERFUL TOOL FOR NANO-PROCESSING—CONTROL OF NUCLEATION AND PARTICLE GROWTH PROCESSES

Recently, nano-particles have drawn great attention due to their excellent characteristics. A microchannel is very attractive to produce nano-particles, as reaction conditions can be precisely controlled quite easily during continuous production. From this viewpoint, a number of studies on forming inorganic and organic fine particles have been conducted. The formation of cadmium selenide fine particles was studied by utilizing a simple microreaction system which consisted of syringe pumps, a capillary tube and an oil bath (Nakamura et al., 2002). It was shown that single-nano-particles can be formed and that the sizes of nano-particles can be precisely controlled by reaction temperature. The production of multi-layer nano-particles with core-shell structure was also examined (Wang et al., 2004). Cadmium selenide was coated by zinc sulfide using a multi-stage microreaction system, and it was shown that the fluorescence efficiency of the prepared particles was higher that of non-coated cadmium selenide particles.

Pigment fine particles were produced by utilizing a multi-lamination type microreactor (Wille et al., 2004). It was shown that the size of fine particles formed by the microreactor was smaller than those obtained using the conventional batch method. The mixing performance for producing polystyrene fine particles was also examined by using a new micromixer (Nagasawa et al., 2005). It was clarified that the particle size and its distribution could be controlled to, respectively, become smaller and narrower by instant mixing based on collision of microsegments. Thus, it is shown that microreactor is a promising tool for improving particle properties through fast mixing and precise temperature controlling. Especially, mixing is the most crucial factor to control nuclei formation and particle growth. In microreaction technology, mixing is classified into two categories: one is precise mixing by controlled diffusion and the other is instant mixing.

In a previous study (Nagasawa and Mae, 2006), a microreactor with same axle dual pipe was developed on the basis of the following concepts: two immiscible liquids were flowed in the inner and outer tubes, respectively, and maintained an annular and laminar flow of separated phase to create a micro space by the outer fluid wall as shown in Fig. 4. In this method, a nucleation section and a particle growth section can be sequentially connected along the flow in the reactor. Mono-modal spherical particles of titania with narrow size distribution were successfully produced without precipitation of particles at the wall. By changing the diameter of inner tube, the particle size was precisely controlled as shown in

Fig. 4. The mean particle size was 45 nm for the tube of 307 μm i.d., 84 nm for the tube of 607 μm i.d., and 121 nm for the tube of 877 μm i.d., respectively. In this system, nuclei formation and particle growth proceed at the interface of two fluids as shown in Fig. 5. In the nuclei formation zone, nuclei are generated as H_2O diffuses to the interface between fluids at the initial part of the microreactor. In the particle growth zone, on the other hand, particles grow uniformly using the ordered TTIP flux which reaches the interface. It is considered that the size distribution depends on the length of the nucleation zone. A method to determine the length of nucleation zone, z0, was presented (Tsujiuchi et al., 2006). Fig. 5 shows the length of the nucleation zone required to achieve a certain particle size distribution using a double same axle microreactor. The figure shows that nuclei formation should be completed at least at the position 5.0×10^{-3} m from the inlet. As detailed and quantitative information on nucleation has been obtained, the fluxes of H_2O and TTIP into the interface can be controlled by changing the shape of microchannel and adjusting its size. In this case, the reduction of inner tube diameter and the designing of the contact patterns of fluids in the nucleation zone is an effective way to obtain particles with a sharp size distribution. I

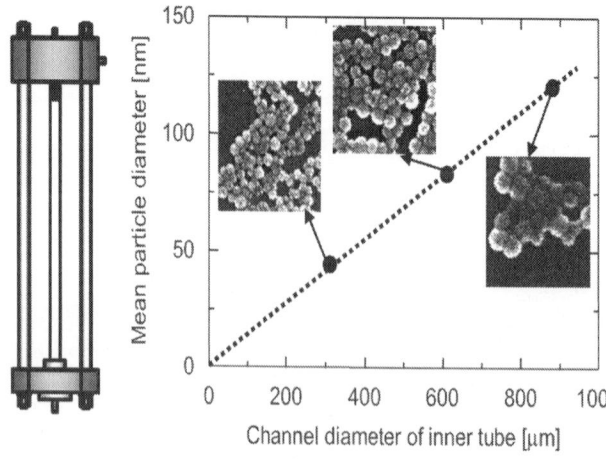

Figure 4: Control of nano-particle size by the size of inner tube.

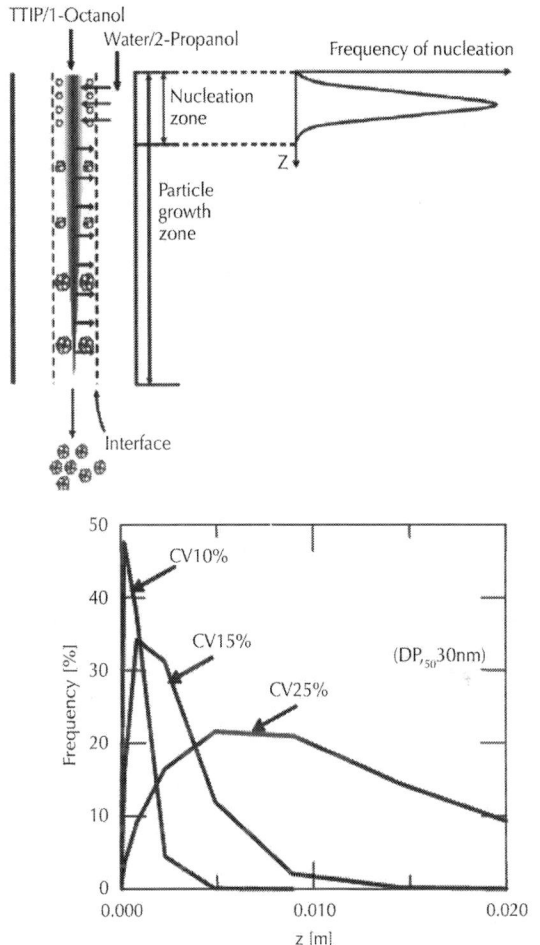

Figure 5: Design of nuclei formation using a annular flow in microchannel.

Instant mixing is also effective for controlling nuclei formation. The K-M mixer (Nagasawa et al., 2005 and Aoki and Mae, 2006) was applied to the production of organic nano-pigments (Maeta et al., 2006). A flowing solution of an organic pigment dissolved in an alkaline aqueous organic solvent is mixed with a precipitation medium in a microchannel. As shown in Fig. 6, a clear and stable dispersion of nano-particles having a narrow size distribution around 20 nm was successfully produced. This was brought about by instant mixing with rapid pH change. Thus, the microdevice provides us a new method to achieve uniform nucli formation during particle synthesis.

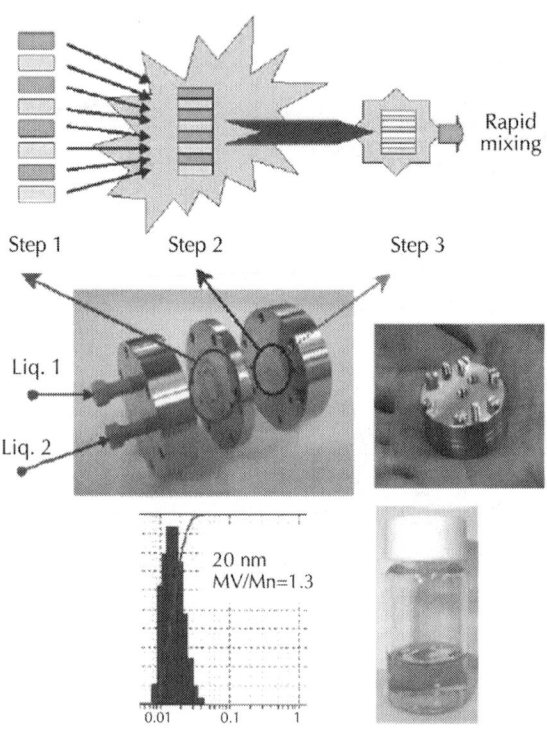

Figure 6: Production nano-organic pigment using K-M mixer.

NEW UNIT OPERATION BY UTILIZ-ING A FUNCTIONAL WALL IN THE MICROCHANNEL

In a microspace, the interaction between fluids and wall is not negligible. This leads to a high pressure loss which is a typical negative feature of microprocessing. However, by taking advantage of this interaction, the bulk properties of fluids can be controlled. As an example, using a micromixer based on repeated splitting and recombination (Mae et al., 2004), a soap-free emulsion was successfully produced at a short contact time of 0.1 s. The maximum production capacity of a single mixer was 200 t/year. By examining the effect of wall type on the size of emulsion, it was clarified that the emulsion was formed by the charge of oil droplets induced by the friction between wall and fluids in microspace as shown in Fig. 7. If this mixing unit is attached before a polymerization reactor, a soap-free polymer product can be easily produced.

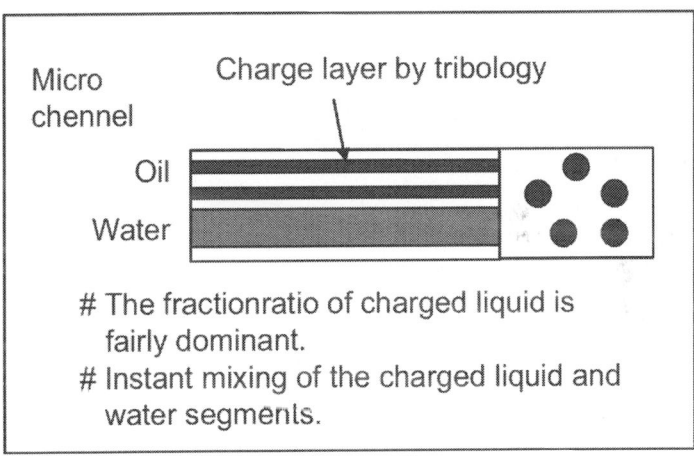

Figure 7: Presumed mechanism for the formation of soap-free emulsion.

A new microchannel device to assist the coalescence of dispersed droplets was developed (Okubo et al., 2004). The microchannel of this device was formed by two flat plates made of glass and PTFE, so its cross-section was rectangular. This device uses the following features: (1) the dispersed droplets are deformed in the planar microchannel with rectangular cross-section and the liquid–liquid interface destabilizes; and (2) the interaction between the PTFE wall and the organic phase has a big influence on the bulk flow in the channel at microscale, and a velocity difference between the continuous and dispersed phases is brought about.

A stable liquid–liquid dispersion made from heptane and hexane as the dispersed phase and a 1.0 wt% sodium dodecyl sulfate aqueous solution as the continuous phase was supplied to this microchannel device (channel depth: 5μm) at a rate of 0.3 cc/min. As shown in Fig. 8, good liquid–liquid separation was attained continuously even though the contact time was shorter than 0.1 s and visible dispersed droplets barely existed in the exit liquid. This proposed device is expected to contribute to rapid efficiency improvement in industrial extraction operations. It can also handle a large amount of dispersion, so it can be used as, for example, an attachment for macroscale equipment. Thus, wall designing could lead to the development of a new unit operation which can be used as a functional attachment of a conventional macro process.

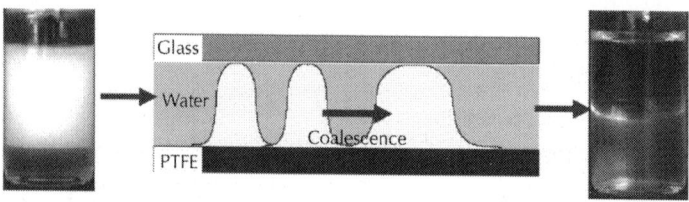

Figure 8: Instant coalescence of emulsion by utilizing the interactions between walls and fluids.

PARADIGM SHIFT TO ADVANCED CHEMICAL REACTION ENGINEERING

As described above, many distinguished results have been obtained using microreactors. In conventional chemical engineering, the design equation for a reactor is to estimate the relationship between the mean conversion and the whole volume of the reactor. Only few models which estimate the performance of macroreactors take molecular diffusion of reactants into account (Dang and Steinberg,

1980, Clifford et al., 1998, Ou and Ranz, 1983 and Chella and Ottino, 1984). However, there is no quantitative equation for designing the details of a reactor, especially its shape. It is, therefore, essential to build up precise models for this new production technology by process intensification. From the above discussions, we can understand that the designing of microfluid segments is the key point to use microdevices to realize unique features which only occur in microspace. From this viewpoint, a concept of microfluid functional segment as a factor to design and operate a microreactor was proposed (Aoki et al., 2004). A microfluid segment is defined as a minimum unit having microproperties which can be used to improve various unit operations and reactions in a microflow. To realize this concept, it is essential to build up a method to both design and operate a microdevice. They are now challenging to present a new model which can be used for this purpose, which takes account of reactor shape by introducing the concept of microfluid segments.

Two dimensionless numbers that represent effects of geometric design factors of fluid segments on reactor performance in reactors using mixing operation have been introduced: the ratio of reaction rate to diffusion rate φ, and the aspect ratio of the mean diffusion lengths in the two-dimensional directions of the reactor cross-section w (Aoki et al., 2006). The geometric design factors of fluid segments applied in this paper are the arrangement and the cross-sectional shape of fluid segments in the reactor inlet. Methods to determine the dimensionless numbers have also been introduced. These numbers can be used to predict reactor performance regardless of geometric design factors, and to determine geometric design factors of fluid segments to obtain a desired reactor performance. To investigate the effectiveness of these numbers, the product yields of reactors having different arrangements and shapes of fluid segments but the same dimensionless numbers were compared using CFD simulations as shown in Fig. 9.

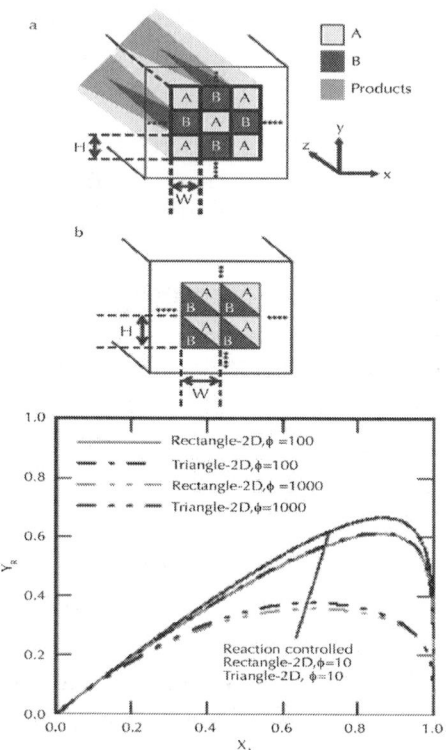

Figure 9: Relations between φ and $_{YR}$ for fluid segments having rectangular and triangular cross-sectional shapes. $C_{B0}/C_{A0}=2; k_2/k_1=0.2$; w=1.

The results show that the order of φ allows us to determine whether reactions proceed under reaction controlled conditions, regardless of the cross-sectional shape of fluid segments. We can determine the largest size of fluid segments to achieve ideal mixing for reaction controlled conditions using the threshold value of φ. The threshold value of φ for reaction controlled conditions depends on the ratio of rate constants of multiple reactions. The second dimensionless number w can represent effects of arrangements and aspect ratios of cross-sectional shapes of fluid segments on the reactor performance. Product yields in reactors whose cross-sectional is a right-angled triangle shape of fluid segments have larger dependence on w than that of a rectangle. When we apply a right-angled triangle to the cross-sectional shape of fluid segments, the fluid segments need to be designed so that w is close to unity. From the concentration profile of reactants in the reactors, reactors whose cross-sectional shape of fluid segments is rectangle are favorable from the viewpoint that uniform mixing in the cross-sectional plane of reactor can be achieved especially when w takes large values.

Another quantitative model was presented for the precise controlling of a segmented flow (Tanthapanichakoon et al., 2006). As mentioned above, a segmented flow is very attractive as rapid mixing can be achieved by the friction of slugs and wall, as residence time can be strictly controlled, and so on. To utilize these distinct features, the quantitative relationships among mixing rate, operation conditions and device dimensions were examined. The modified Peclet number, $Pe^* = U_s d_s^2 / ID$, which indicates the ratio of diffusion rate to circulating rate driven by the friction between wall and slug, was presented for designing mixing in liquid slugs. Based on a new parameter, the relationship between mixing rate and the modified Peclet number was established to aid the determining of reactor size and operation conditions. Thus, chemical reaction engineering has entered a new era in which precise description of a reactor including its shape and arrangement can be presented. It is expected that many models will emerge in the near future, and a new version of chemical reaction engineering will be established.

FOR THE FUTURE

Research on microreactors has developed in Germany from the 1990s. Now, several pilot plants using microreactors are being operated throughout the world. However, the question "Can microreaction technology be applied to industrial mass production?" always comes up. Indeed, it is difficult, if not impossible, to completely replace bulk processes which are being operated at million tons per year scale with microreaction systems, but such systems may be used to provide new functions in plant retrofitting. On the other hand, as profits in chemical industries are mainly obtained through the production of fine chemicals using batch processes, microreaction technology should be actively applied to such areas. The key point is to shift to more severe reaction conditions and achieve high throughputs by shortening residence time. The first step in the usage of microreaction technology is to partly introduce microdevices to existing macroprocesses, and utilize their unique functions such as rapid mixing and quenching. Many technical barriers ought to be overcome for such integration and critical cost analysis should be conducted, but microreaction technology is sure to evolve and become a new environmentally benign processing technology, which will be widely used throughout the world within a decade.

ACKNOWLEDGMENTS

I appreciate very much to Prof. Shin R. Mukai and Dr. Nobuaki Aoki for the fully discussion and the preparation of graphs, respectively.

REFERENCES

1. Aoki, N., Hasebe, S., Mae, K., 2004. Mixing in microreactors: effectiveness of lamination segments as a form of feed on product distribution for multiple reactions. Chemical Engineering Journal 101, 323–331.

2. Aoki, N., Mae, K., 2006. Effects of channel geometry on mixing performance of micromixers using collision of fluid segments. Chemical Engineering Journal 118, 189–197.

3. Aoki, N., Hasebe, S., Mae, K., 2006. Geometric design of fluid segments in microreactors using dimensionless numbers. A.I.Ch.E. J. 52 (4), 1502–1515.

4. Auroux, P.-A., Iossifidis, D., Reyes, D.R., Manz, A., 2002. Micro total analysis systems. 2. Analytical standard operations and applications. Analytical Chemistry 74, 2637–2652.

5. Brivio, M., Oosterbroek, R.E., Verboom, W., Goedbloed, M.H., van den Berg, A., Reinhoudt, D.N., 2003. Surface effects in the esterification of 9- pyrenebutyric acid within a glass microreactor. Chemical Communications 8, 1924–1925.

6. Burns, J.C., Ramshaw, C., 2001. The intensification of rapid reactions in multiphase systems using slug flow in capillaries. Lab on a Chip 1, 10–15.

7. Chambers, R.D., Spink, R.C.H., 1999. Microreactor for elemental fluorine. Chemical Communications, 883–884.

8. Chella, R., Ottino, J.M., 1984. Conversion and selectivity modifications due to mixing in unpremixed reactors. Chemical Engineering Science 39, 551–567.

9. Clifford, M.J., Roberts, E.P.L., Cox, S.M., 1998. The influence of segregation on the yield for a series-parallel reaction. Chemical Engineering Science 53, 1791–1801.

10. Daito, N., Aoki, N., Yoshida, J., Mae, K., 2006. Selective condensation reaction of phenols and formaldehyde using micromixers based on collision of fluid segments. Industrial and Engineering Chemistry Research 45, 4954–4961.

11. Dang, V., Steinberg, M., 1980. Convective diffusion with homogeneous and heterogeneous reactions in a tube. Journal of Physical Chemistry 84, 214–219.

12. De Mello, A., Wootton, R., 2002. But what is it good for? Applications of microreactor technology for the fine chemical industry. Lab on a Chip 2, 7N–13N.

13. Dummann, G., Quitmann, U., Groschel, L., Agar, D.W., Morgenschweis, K., 2003. The capillary-microreactor: a new reactor concept for the intensification of heat and mass transfer in liquid–liquid reactions. Catalysis Today 79–80, 433–439.

14. Ehrfeld, W., Golbig, K., Hessel, V., Löwe, H., Richter, T., 1999. Characterization of mixing in micromixers by a test reaction: single mixing units and mixer arrays. Industrial and Engineering Chemistry Research 38, 1075–1082.

15. Ehrfeld, W., Hessel, V., Löwe, H., 2000. Microreactors. Wiley-VCH, Weinheim.

16. Engler, M., Kockmann, N., Kiefer, T., Woias, P., 2004. Numerical and experimental investigations on liquid mixing in static micromixers. Chemical Engineering Journal 101, 315–322.

17. Fletcher, P.D.I., Haswell, S.J., Pombo-Villar, E., Warrington, B.H., Watts, P., Wong, S.Y.F., Zhang, X., 2002. Microreactors: principles and applications in organic synthesis. Tetrahedron 58, 4735–4757.

18. Fukuyama, T., Shinmen, M., Nishitani, S., Sato, M., Ryu, I., 2002. A copperfree Sonogashira coupling reaction in ionic liquids and its application to a microflow system for efficient catalyst recycling. Organic Letters 4, 1691–1694.

19. Geschke, O., Klank, H., Tellemann, P., 2004. Microsystem Engineering of Lab-on-a-Chip Devices. Wiley-VCH, Weinheim.

20. Gobby, D., Angeli, P., Gavriilidis, A., 2001. Mixing characteristics of T-type microfluidic mixers. Journal of Micromechanics and Microengineering 11, 126–132.

21. Harrison, D.J., Fluri, K., Seiler, K., Fan, Z., Effenhauser, C.S., Manz, A., 1993. Micromachining a miniaturized capillary electrophoresis-based chemical analysis system on a chip. Science 261, 895–897.

22. Haswell, S.J., Middleton, R.J., O'Sullivan, B., Skelton, V., Watts, P., Styring, P., 2001. The application of microreactors to synthetic chemistry. Chemical Communications, 391–398.

23. Haverkamp, V., Emig, G., Hessel, V., Liauw, M., Löwe, H., 2001. Characterization of a gas/liquid microreactor, the microbubble column: determination of specific interfacial area. In: Proceedings of the Fifth International Conference on Microreaction Technology, pp. 202–214.

24. Henkel, T., Bermig, M., Kielpinski, M., Grodian, A., Metze, J., Kohler, J.M., 2004. Chip modules for generation and

manipulation of fluid segments for micro serial flow process. Chemical Engineering Journal 101, 439–445.

25. Herweck, T., Hardt, S., Hessel, V., Löwe, H., Hofmann, C., Weise, F., Dietrich, T., Freitag, A., 2001. Visualization of flow patterns and chemical synthesis in transparent Micromixers. In: Proceedings of the Fifth International Conference on Microreaction Technology, pp. 215–230.

26. Hessel, V., Ehrfeld, W., Golbig, K., Haverkamp, V., Löwe, H., Storz, M., Wille, Ch., Guber, A.E., Jahnisch, K., Baerns, M., 2000. Gas/liquid microreactors for direct fluorination of aromatic compounds using elemental fluorine. In: Proceedings of the Third International Conference on Microreaction Technology, pp. 526–540.

27. Hessel, V., Hardt, S., Löwe, H., Schönfeld, F., 2003. Laminar mixing in different interdigital micromixers: I. Experimental characterization. AIChE Journal 49, 566–577.

28. Hessel, V., Hardt, S., Löwe, H., 2004a. Chemical Microprocess Engineering. WILEY-VCH, Weinheim, Germany.

29. Hessel, V., Hofmann, C., Löwe, H., Meudt, A., Scherer, S., Schönfeld, F., Werner, B., 2004b. Selectivity gains and energy savings for the industrial phenyl boronic acid process using micromixer/tubular reactors. Organic Process Research and Development 8, 511–523.

30. Hessel, V., Löwe, H., 2005. Microchannel engineering: components, plant concepts, user acceptance—part III. Chemical Engineering Technology 28 (5), 531–544.

31. Hessel, V., Löwe, H., Schönfeld, F., 2005a. Micromixers—a review on passive and active mixing principles. Chemical Engineering Science 60, 2479–2501.

32. Hessel, V., Hofmann, C., Löb, P., Löhndorf, J., Löwe, H., Ziogas, A., 2005b. Aqueous Kolbe–Schmitt synthesis using resorcinol in a microreactor laboratory rig under high-p,T conditions. Organic Process Research and Development 9 (4), 479–489.

33. Iwasaki, T., Naoya Kawano, N., Yoshida, J., 2006. Radical polymerization using microflow system: numbering-up of microreactors and continuous operation. Organic Process Research and Development 10 (6), 1126–1131.

34. Jensen, K.F., 2001. Microreaction engineering—is smaller better? Chemical Engineering Science 56, 293–303.

35. John, M.K., Kim, J.-H., Noh, J., Woo, S.I., Yoon, E., Park, H.G., 2003. Design of a recycle micromixer. In: Proceedings of the Seventh International Conference on m-TAS, pp. 109–112.

36. Kawaguchi, T., Miyata, H., Ataka, K., Mae, K., Yoshida, J., 2005. Room temperature swern oxidation using micro flow system. Angewandte Chemie International Edition 44, 2413–2416.

37. Löb, P., Drese, K.S., Hessel, V., Hardt, S., Hofmann, C., Löwe, H., Schenk, Y., Schönfeld, F., Werner, B., 2004. Steering of liquid mixing speed in interdigital micromixers—from very fast to deliberately slow mixing. Chemical Engineering Technology 27, 340–345.

38. Losey, M.W., Schmnidt, M.A., Jensen, K.F., 2001. Microfabricated multiphase packed-bed reactors: characterization of mass transfer and reactions. Industrial and Engineering Chemistry Research 40, 2555–2562.

39. Mae, K., Maki, T., Hasegawa, I., Eto, U., Mizutani, Y., Honda, N., 2004. Development of a new micromixer based on split/recombination for mass production and its application to soap free emulsifier. Chemical Engineering Journal 101, 31–38.

40. Maeta, H., Sato, T., Nagasawa, H., Mae, K., 2006. New synthetic method of organic pigment nano-particle by microreactor system. In: Proceedings of the A.I.Ch.E. Spring Meeting, Topical 1.

41. Maki, T., Ueyama, T., Mae, K., 2005. Methanol decomposition by use of assemble-type microreactor. Chemical Engineering Technology 28 (4), 494–500.

42. Mohr, W.D., Saxton, R.L., Jepson, C.H., 1957. Mixing in laminar-flow systems. Industrial and Engineering Chemistry 49, 1855–1856.

43. Nagaki, A., Togai, M., Suga, S., Aoki, N., Mae, K., Yoshida, J., 2005. Control of extremely fast competitive consecutive reactions using micromixing. Selective Friedel–Crafts aminoalkylation. Journal of the American Chemical Society 127, 11666–11675.

44. Nagasawa, H., Mae, K., 2006. Design of a new micromixer for instant mixing based on annular microsegments for fine particle production. Industrial and Engineering Chemistry Research 45 (7), 2179–2186.

45. Nagasawa, H., Aoki, N., Mae, K., 2005. Design of a new micromixer for instant mixing based on collision of microsegment. Chemical Engineering Technology 28, 324–330.

46. Nakamura, H., Yamaguchi, Y., Miyazaki, M., Maeda, H., Uehara, M., Mulvaney, P., 2002. Preparation of CdSe nanocrystals in a micro-flowreactor. Chemical Communications 23, 2844–2845.

47. Okubo, Y., Toma, M., Ueda, H., Maki, T., Mae, K., 2004. Microchannel devices for the coalescence of dispersed droplets produced for use in rapid extraction processes. Chemical Engineering Journal 101 (1–3), 39–48.

48. Ookawara, S., Minamimoto, K., Ogawa, K., 2004. Stability of interface between two liquids in T-shape confluence of microchannels. Kagaku Kogaku Ronbunshu 30, 148–153.

49. Ou, J., Ranz, W.E., 1983. Mixing and chemical reactions: a contrast between fast and slow reactions. Chemical Engineering Science 38, 1005–1013.

50. Reyes, D.R., Iossifidis, D., Auroux, P.-A., Manz, A., 2002. Micro total analysis systems. 1. Introduction, theory, and technology. Analytical Chemistry 74, 2623–2636.

51. Salimi-Moosavi, H., Tang, T., Harrison, D.J., 1997. Electroosmotic pumping of organic solvents and reagents

in microfabricated reactor chips. Journal of the American Chemical Society 119, 8716–8717.

52. Sato, O., Ikushima, Y., Yokoyama, T., 1998. Noncatalytic Beckmann rearrangement of cyclohexanone-oxime in supercritical water. Journal of Organic Chemistry 63, 9100–9102.

53. Schwalbe, T., Kursawe, A., Sommer, J., 2005. Application report on operating cellular process chemistry plants in fine chemicals and contact manufacuturing industries. Chemical Engineering Technology 28 (4), 408–419.

54. Skelton, V., Greenway, G.M., Haswell, S.J., Styring, P., Morgan, D.O., 2000. Microreactor synthesis: synthesis of cyanobiphenyls using a modified Suzuki coupling of an aryl halide and aryl boronic acid. In: Proceedings of the Third International Conference on Microreaction Technology, pp. 235–242.

55. Song, H., Bringer, M.R., Tice, J.D., Gerdts, C.J., Ismagilov, R.F., 2003. Experimental test of scaling of mixing by chaotic advection in droplets moving through microfluidic channels. Applied Physics Letters 83, 4664–4666.

56. Suga, S., Nagaki, A., Yoshida, J., 2003a. Highly selective Friedel–Crafts monoalkylation using micromixing. Chemical Communications, 354–355.

57. Suga, S., Nagaki, A., Tsutsui, Y., Yoshida, J., 2003b. N-acyliminium ion pool as hetero diene in [4 + 2] cycloaddition reaction. Organic Letters 5, 945–947.

58. Tanthapanichakoon, W., Aoki, N., Matsuyama, K., Mae, K., 2006. Design of mixing in microfluidic liquid slugs based on a new dimensionless number for precise reaction and mixing operations. Chemical Engineering Science 61, 4220–4232.

59. Thayer, A.M., 2005. Harnessing microreactions. C&EN Coverstory, vol. 83(22), May 30, 2005, pp. 43–52.

60. Tsujiuchi, T., Nagasawa, H., Maki, T., Mae, K., 2006. Control of nuclei formation and aggregation processes for nano-particles using a microreactor with same axle dual pipe.

In: Proceedings of the Ninth International Conference on Microreaction Technology, Topic 4, No. 35.

61. Wang, H., Li, X., Uehara, M., Yamaguchi, Y., Nakamura, H., Miyazaki, M., Shimizu, H., Maeda, H., 2004. Continuous synthesis of CdSe–ZnS composite nano-particles in microfluidic reactor. Chemical Communications 1, 48–49.

62. Wang, Y., Lin, Q., Mukherjee, T., 2005. A model for laminar diffusion-based complex electrokinetic passive micromixers. Lab on a Chip 5, 877–887.

63. Wieûmeier, G., Hönicke, D., 1996. Heterogeneously catalyzed gas-phase hydrogenation of cis,trans,trans-1,5,9-Cyclododecatriene on palladium catalysts having regular pore systems. Industrial and Engineering Chemistry Research 35, 4412–4416.

64. Watts, P., Wiles, C., Haswell, S.J., Pombo-Villar, E., Styring, P., 2001. The synthesis of peptides using microreactors. Chemical Communications, 990–991.

65. Wiles, C., Watts, P., Haswell, S.J., Pombo-Villar, E., 2001. The aldol reaction of silyl enol ethers within a microreactor. Lab on a Chip 1, 100–101.

66. Wiles, C., Watts, P., Haswell, S.J., Pombo-Villar, E., 2002. 1,4-Addition of enolates to α, β-unsaturated ketones within a microreactor. Lab on a Chip 2, 62–64.

67. Wille, Ch., Gabski, H.-P., Haller, Th., Kim, H., Unverdorben, L., Winter, R., 2004. Synthesis of pigments in a three-stage microreactor pilot plant—an experimental technical report. Chemical Engineering Journal 101, 179–184.

68. Wolfrath, O., Kiwi-Minsker, L., Renken, A., 2001. Novel membrane reactor with filamentous catalytic bed for propane Dehydrogenation. Industrial and Engineering Chemistry Research 40, 5234–5239.

69. Wong, S.H., Ward, M.C.L., Wharton, C.W., 2004. Micro T-mixer as a rapid mixing micromixer. Sensors and Actuators B 100, 359–379.

70. Wörz, O., Jäckel, K.P., Richter, T., Wolf, A., 2001. Microreactors, a new efficient tool for optimum reactor design. Chemical Engineering Science 56, 1029–1033.

71. Yang, R., Williams, J.D., Wang, W., 2004. A rapid micromixer/reactor based on arrays of spatially impinging micro-jets. Journal of Micromechanics and Microengineering 14, 1345–1351.

72. Yube, K., Mae, K., 2005. Efficient oxidation of aromatics with peroxides under severe conditions using a microreaction system. Chemical Engineering Technology 28 (3), 331–336.

73. Zheng, B., Tice, J.D., Ismagilov, R.F., 2003. Formation of droplets and mixing in multiphase microfluidics at low Reynolds and Capillary numbers. Langmuir 19, 9127–9133.

74. Zheng, B., Tice, J.D., Ismagilov, R.F., 2004. Formation of droplets of alternating composition in microfluidic channels and applications to indexing of concentrations in droplet-based assays. Analytical Chemistry 76, 4977–4982.

Chapter 6

Heat and Mass Transfer in Unsteady Rotating Fluid Flow with Binary Chemical Reaction and Activation Energy

Faiz G. Awad, Sandile Motsa, and
Melusi Khumalo

[1]Department of Pure & Applied Mathematics, University of Johannesburg, Auckland Park, Johannesburg, South Africa

[2]School of Mathematics, Statistics and Computer Science, University of KwaZulu-Natal, Scottsville, Pietermaritzburg, South Africa

ABSTRACT

In this study, the Spectral Relaxation Method (SRM) is used to solve the coupled highly nonlinear system of partial differential equations due to an unsteady flow over a stretching surface in an incompressible rotating viscous fluid in presence of binary chemical reaction and Arrhenius activation energy. The velocity, temperature and concentration distributions as well as the skin-friction, heat and mass transfer coefficients have been obtained and discussed for various physical parametric values. The numerical results obtained by (SRM) are then presented graphically and discussed to highlight the physical implications of the simulations.

INTRODUCTION

The study of boundary layer flow and heat transfer inducted by stretching surface has attracted considerable interest due to its wide applications in industrial processes such as the cooling of an infinite metallic plate in a cooling bath, the aerodynamic extrusion of plastic sheets, boundary layer along the material handling conveyers, the boundary layer along a liquid film and condensation processes. The quality of the final product depends on the skin friction coefficient and the rate of heat transfer. One of the earliest studies of the boundary layer flow problem was conducted by Sakiadis [1], [2]. Crane [3] extended this concept to present the problem of the steady two-dimensional boundary layer flow over stretching sheet of elastic flat surface with linear velocity. He demonstrated that the problem was interesting because it possessed a closed form exact solution. Studies have been carried out for the case of the axisymmetric and three-dimensional flow by Brady and Acrivos [4], and Wang [5]. Investigations by, among others, Afzal [6], Prasad et al. [7], Abel and Mahesha [8], Bataller [9], Abel et al. [10], have also provided examples of various aspects of this important field.

Unsteady flows in rotating fluid have numerous uses or potential applications in chemical and geophysical fluid dynamics and mechanical nuclear engineering. Using the Fourier series analysis, Soundalgekar et al. [11] investigated the unsteady rotating flow of incompressible, viscous fluid past an infinite porous plate. The boundary layer flow problem formed in a rotating fluid by oscillating flow over an infinite half-plate has been examined Bergstrom [12]. Abbas et al. [13] studied the unsteady boundary layer MHD flow and heat transfer on a stretching continuous sheet in a viscous incompressible rotating fluid numerically using the Keller-box method. Nazar et al. [14] investigated unsteady flow due to the impulsive starting from rest of a stretching surface in a viscous and incompressible rotating fluid. Zheng et al. [15] studied the unsteady rotating flow of a generalized Maxwell fluid with fractional derivative model between two infinite straight circular cylinders. Using the shooting method Fang [16] studied the problem of the laminar unsteady flow over a stretchable rotating disk with deceleration is investigated. Rashad [17] investigated the unsteady magnetohydrodynamics boundary-layer flow and heat transfer for a viscous laminar incompressible electrically conducting and rotating fluid due to a stretching surface embedded in a saturated porous medium with a temperature-dependent viscosity in the presence of a magnetic field and thermal radiation effects. Nageeb et al. [18]used the Runge-Kutta method based on shooting technique to investigate the unsteady MHD flow and heat transfer of a couple stress fluid over a rotating disk. For the case in which steady flow rotating flow involve the powe-law, very recently, Hajmohammadi et al. [19] developed an analytical solution for two-phase flow betwen two rotating cylinders filed with power law liquid and a micro layer of gas. Moreover Hajmohammadi and Nourazar [20] the problem of heat transfer repercussions thin gas layer in micro cylindrical Couette flows involving power-law liquids.

Many chemically reacting systems involve the species chemical reactions with finite Arrhenius activation energy, with examples occurring in geothermal and oil reservoir engineering. The interactions between mass transport and chemical reactions are generally very complex, and can be observed in the production and consumption of reactant species at different rates both within the fluid and the mass transfer. One of the earliest studies involving the binary chemical reaction in boundary layer flow was published by Bestman [21] who presented an analytical solution using the perturbation method to show the effect of the activation energy in natural convection in a porous medium. Using the Arrhenius activation energy Bestman [22]subsequently investigated radiative heat transfer on the flow of a combustible mixture in a vertical pipe. Makinde et al. [23] studied the effects of nth order Arrhenius chemical reaction, thermal radiation, suction/injection and buoyancy forces on unsteady convection of a viscous incompressible fluid past a vertical porous plate numerically. They showed that the effect of the chemical reaction, heat source, and suction or injection is significant at the wall of the wedge on the flow field. A numerical study of the unsteady mixed convection with Dufour and Soret effects past a semi-infinite vertical porous flat plate moving through a binary mixture of chemically reacting fluid was conducted by Makinde and Olanrewaju [24]. The most recent contributions in this area include those of Abdul Maleque [25]–[27], who investigated the effects of chemical reactions with Arrhenius activation Energy on unsteady convection heat and mass transfer boundary layer fluid flow.

This work deals with the effects of chemical reactions with finite Arrhenius activation energy on unsteady rotating fluid flow due to a stretching surface with Binary chemical reaction and activation energy. The governing partial differential equations are solved using the spectral relaxation method (SRM). The SRM is based on simple decoupling and rearrangement of the governing nonlinear equations in a Gauss-Seidel manner. The resulting sequence of equations are integrated using the Chebyshev spectral collocation method. The SRM was introduced in[29] for the solution of the nonlinear ODE system model of von Karman flow of a Reiner-Rivlin fluid. A generalised presentation of the method was later presented in [30] and applied in three ODE based systems of boundary layer flow equations of varying complexity. The method has also been successfully used in the solution of chaotic and hyper-chaotic systems [31], [32]which are defined as systems of ODE initial value problems.

Mathematical Formulation

Consider the three-dimensional, unsteady flow due to a stretching surface in a rotating fluid. The motion in the fluid is three dimensional. At time t~0, the surface z~0 is impulsively stretched in the x direction in the rotating fluid. The velocity components are assume to be (u,v,w) in the direction of the Cartesian axes (x,y,z), respectively, and the axes is rotating at an angular velocity in the z direction. The surface temperature T_w and solute concentration C_w are higher than the ambient values T_∞ and C_w, respectively. Assuming a species chemical reaction with finite Arrhenius activation energy, the governing equations for the problem can be written in the form

$$\frac{\partial u}{\partial x} + \frac{\partial v}{\partial y} + \frac{\partial w}{\partial z} = 0,$$

(1)

$$\frac{\partial u}{\partial t} + u\frac{\partial u}{\partial x} + v\frac{\partial u}{\partial y} + w\frac{\partial u}{\partial z} - 2\Omega v = -\frac{1}{\rho}\frac{\partial p}{\partial x} + \nu\nabla^2 u,$$

(2)

$$\frac{\partial v}{\partial t} + u\frac{\partial v}{\partial x} + v\frac{\partial v}{\partial y} + w\frac{\partial v}{\partial z} + 2\Omega v = -\frac{1}{\rho}\frac{\partial p}{\partial x} + \nu\nabla^2 v,$$

(3)

$$\frac{\partial w}{\partial t} + u\frac{\partial w}{\partial x} + v\frac{\partial w}{\partial y} + w\frac{\partial w}{\partial z} = -\frac{1}{\rho}\frac{\partial p}{\partial z} + \nu\nabla^2 w,$$

(4)

$$\frac{\partial T}{\partial t} + u\frac{\partial T}{\partial x} + v\frac{\partial T}{\partial y} + w\frac{\partial T}{\partial z} = \alpha\nabla^2 T,$$

(5)

$$\frac{\partial C}{\partial t} + u\frac{\partial C}{\partial x} + v\frac{\partial C}{\partial y} + w\frac{\partial C}{\partial z} =$$

$$D\nabla^2 C - k_r^2\left(\frac{T}{T_\infty}\right)^n e^{-\frac{E_a}{\kappa T}}(C - C_\infty),$$

(6)

where p is the pressure, p is the density, v is the kinematic viscosity, T is the fluid temperature, C is the solutal concentration, α is the thermal diffusivity, D is the solutal diffusivity and ∇^2 denotes the three-dimensional Laplacian, $\left(\frac{T}{T\infty}\right)^n \exp\left(E_a / \{kT\}\right)$ is the modified Arrhenius function, k is the Boltzmann constant, k_r^2 is the chemical reaction rate constant, n is a unit less constant exponent fitted rate constants typically lie in the range -1<n<1. Let the surface be impulsively stretched in the x direction such that the initial and boundary conditions are

$$t \geq 0 : u = ax, v = 0, w = 0, T = T_w, C = C_w, \quad \text{at} \quad z = 0,$$

$$u \to 0, w \to 0, \quad T \to T_\infty \quad C \to C_\infty \quad \text{as} \quad z \to \infty, \tag{7}$$

$$t < 0 : u = 0, v = 0, w = 0, T = 0, C = 0 \text{ for all } x, y, z. \tag{8}$$

The following non-dimensional variables are introduced,

$$\eta = \sqrt{\frac{a}{\nu \xi}} z, \xi = 1 - \exp(-\tau), \tau = at, u = axf'(\xi, \eta), v = axh(\xi, \eta),$$

$$w = -\sqrt{a\nu \xi} f(\xi, \eta), \theta(\xi, \eta) = \frac{T - T_\infty}{T_w - T_\infty}, \phi(\xi, \eta) = \frac{C - C_\infty}{C_w - C_\infty}. \tag{9}$$

The governing equations (2) – (5) along with the boundary conditions (7) can be presented as

$$f''' + (1 - \xi)\frac{\eta}{2}f'' + \xi[ff'' - f'^2 + 2\lambda h] = \xi(1 - \xi)\frac{\partial f'}{\partial \xi}, \tag{10}$$

$$h'' + (1 - \xi)\frac{\eta}{2}h' + \xi[fh' - f'h - 2\lambda f'] = \xi(1 - \xi)\frac{\partial h}{\partial \xi}, \tag{11}$$

$$\frac{1}{Pr}\theta'' + (1 - \xi)\frac{\eta}{2}\theta' + \xi f\theta' = \xi(1 - \xi)\frac{\partial \theta}{\partial \xi}, \tag{12}$$

$$\frac{1}{Sc}\phi'' + (1 - \xi)\frac{\eta}{2}\phi' + \xi f\phi' - \sigma^2 \xi\phi(1 + n\delta\theta)\exp\left(-\frac{E}{1 + \delta\theta}\right) =$$

$$\xi(1 - \xi)\frac{\partial \phi}{\partial \xi}, \tag{13}$$

subject to the boundary conditions

$$f'(\xi, 0) = 1, f(\xi, 0) = 0, h(\xi, 0) = 0, \theta(\xi, 0) = 1, \phi(\xi, 0) = 1, \xi \geq 0,$$

$$f'(\xi, \infty) \to 0, h(\xi, \infty) \to 0, \theta(\xi, \infty) \to 0, \phi(\xi, \infty) \to 0, \xi \geq 0, \tag{14}$$

where $\lambda = \Omega/a$ is the rotation rate parameter, $Pr = \frac{\nu}{\alpha}$ is the Prandtl number, Sc=v/D is the Schmidt number, E=E$_a$/(kT∞) the non-dimensional activation energy, $\delta = \left(T_w - T_\infty\right) / T_\infty$ is the temperature

relative parameter, $\sigma = \dfrac{kr^2}{\alpha}$ is the dimensionless chemical reaction rate constant.

The non-dimensional skin friction in both x and y directions, the local Nusselt number, the local Sherwood number are defined in the form

$$C_f^x = \frac{\tau_w^x}{\rho(ax)^2}, \quad C_f^y = \frac{\tau_w^y}{\rho(ax)^2}, Nu_x =$$

$$\frac{-x}{T_w - T_\infty}\left(\frac{\partial T}{\partial z}\right)\bigg|_{z=0}, Sh_x = \frac{-x}{C_w - C_\infty}\left(\frac{\partial C}{\partial z}\right)\bigg|_{z=0} \quad (15)$$

where the wall shear stresses τ_w^x, τ_w^y, respectively, are given by

$$\tau_w^x = \mu \frac{\partial u}{\partial z}\bigg|_{z=0}, \qquad \tau_w^y = \mu \frac{\partial v}{\partial z}\bigg|_{z=0} \qquad (16)$$

substituting (9) and (16) into (15) it gives

$$Re_x^{\frac{1}{2}} C_f^x = \xi^{-\frac{1}{2}} \frac{\partial^2 f}{\partial \eta^2}\bigg|_{\eta=0}, \quad Re_x^{\frac{1}{2}} C_f^y = \xi^{-\frac{1}{2}} \frac{\partial h}{\partial \eta}\bigg|_{\eta=0},$$

$$Re_x^{-\frac{1}{2}} Nu_x = \xi^{-\frac{1}{2}} \frac{\partial \theta}{\partial \eta}\bigg|_{\eta=0}, \quad Re_x^{-\frac{1}{2}} Sh_x = \xi^{-\frac{1}{2}} \frac{\partial \phi}{\partial \eta}\bigg|_{\eta=0},$$

where $Re_x = \dfrac{(ax)x}{v}$ is the local Reynolds number.

Numerical Solution

In this section, the spectral relaxation method (SRM) is applied to solve the governing nonlinear PDEs (10–13). For the implementation of the spectral collocation method, at a later stage, it is convenient to reduce the order of equation (10) from three to two. To this end, we set $f' = u$, so that equation (10) becomes

$$u'' + \frac{1}{2}\eta(1-\xi)u' + \xi[fu' - u^2 + 2\lambda h] = \xi(1-\xi)\frac{\partial u}{\partial \xi},$$
(17)

$$f' = u.$$
(18)

The spectral relaxation method algorithm uses the idea of the Gauss-Seidel method to decouple the governing systems of equations (10 – 13). From the decoupled equations an iteration scheme is developed by evaluating linear terms in the current iteration level (denoted by r+1) and nonlinear terms in the previous iteration level (denoted by r). Applying the SRM on (11 – 13) and (17 – 18) gives the following linear partial differential equations;

$$u''_{r+1} + a_{1,r}u'_{r+1} + a_{2,r} = \xi(1-\xi)\frac{\partial u_{r+1}}{\partial \xi},$$
(19)

$$f'_{r+1} = u_{r+1}, \quad f_{r+1}(0,\xi) = 0,$$
(20)

$$h''_{r+1} + b_{1,r}h'_{r+1} + b_{2,r}h_{r+1} + b_{3,r} = \xi(1-\xi)\frac{\partial h_{r+1}}{\partial \xi},$$
(21)

$$\frac{1}{Pr}\theta''_{r+1} + c_{1,r}\theta'_{r+1} = \xi(1-\xi)\frac{\partial \theta_{r+1}}{\partial \xi},$$
(22)

$$\frac{1}{Sc}\phi''_{r+1} + c_{1,r}\phi'_{r+1} + d_{1,r}\phi_{r+1} = \xi(1-\xi)\frac{\partial b_{r+1}}{\partial \xi},$$
(23)

$$u_{r+1}(0,\xi) = 1, \quad u_{r+1}(\infty,\xi) = 0,$$

$$h_{r+1}(0,\xi) = 0, \quad h_{r+1}(\infty,\xi) = 0,$$
(24)

$$\theta_{r+1}(0,\xi) = 1, \quad \phi_{r+1}(0,\xi) = 1,$$
(25)

$$\theta_{r+1}(\infty,\xi) \to 0, \quad \phi_{r+1}(\infty,\xi) \to 0,$$
(26)

Where

$$a_{1,r} = \frac{1}{2}\eta(1-\xi) + \xi f_r, \quad a_{2,r} = -\xi u_r^2 + 2\xi\lambda h_r,$$

$$b_{1,r} = c_{1,r} = \frac{1}{2}\eta(1-\xi) + \xi f_{r+1},$$

$$d_{1,r} = -\lambda_1^2 \xi\phi(1 + n\delta\theta_{r+1})\exp\left(-\frac{E}{1+\delta\theta_{r+1}}\right)$$

The initial approximation for solving equations (10 – 13) are obtained as the solutions at $\xi = 0$. Thus $f_0(\eta,\xi)$, $u_0(\eta,\xi)$, $h_0(\eta,\xi)$, $q_0(\eta,\xi)$ and $b_0(\eta,\xi)$ are given by

$$f_0(\eta,\xi) = \eta\,\mathrm{erfc}\left(\frac{\eta}{2}\right) + \frac{2}{\sqrt{\pi}}\left[1 - \exp\left(-\frac{\eta^2}{4}\right)\right],$$

$$u_0(\eta,\xi) = \mathrm{erfc}\left(\frac{\eta}{2}\right),$$

(27)

$$\theta_0(\eta,\xi) = 1 - \mathrm{erf}\left(\frac{\sqrt{Pr}\,\eta}{2}\right), \quad \phi_0(\eta,\xi) = 1 - \mathrm{erf}\left(\frac{\sqrt{Sc}\,\eta}{2}\right).$$

(28)

Starting from given initial approximations (27 – 28), the iteration schemes (19 – 26) can be solved iteratively for $u_r + 1(\eta,\xi)$, $f_r + 1(\eta,\xi)$, etc, when r=0,1,2,... To solve equation (19 –26) the the linear equations are discretized using the Chebyshev spectral method in the η-direction and use an implicit finite difference method in the ξ-direction. For brevity, the details of the spectral methods are omitted. Interested readers may refer to Refs. [28], [33]. Before applying the spectral method, it is convenient to transform the domain on which the governing equation is defined to the interval [−1,1] where the spectral method can be implemented. For the convenience of the numerical computations, the semi-infinite domain in the space direction is approximated by the truncated domain $[0,\eta_\infty]$, where η_∞ is a finite number selected to be large enough to represent the behaviour of the flow properties when η

is very large. We use the transformation $\eta = \eta_\infty (Y+1)/2$ to map the interval $[0, \eta_\infty]$ to $[-1,1]$. The basic idea behind the spectral collocation method is the introduction of a differentiation matrix D which is used to approximate the derivatives of the unknown variables f, u, h, θ and ϕ at the collocation points (grid points) as the matrix vector product

$$\frac{df}{d\eta}\bigg|_{\eta=\eta_j} = \sum_{k=0}^{N_x} \mathbf{D}_{jk} f(Y_k, \xi) = \mathbf{D}F, \quad j=0,1,\ldots,N_x$$

29)

where $N_x + 1$ is the number of collocation points, $\mathbf{D} = 2D/\eta_\infty$, and

$$F = [f(Y_0,\xi), f(Y_1,\xi), \ldots, f(Y_{N_x},\xi)]^T,$$

$$U = [u(Y_0,\xi), u(Y_1,\xi), \ldots, u(Y_{N_x},\xi)]^T,$$

$$H = [h(Y_0,\xi), h(Y_1,\xi), \ldots, h(Y_{N_x},\xi)]^T,$$

$$Q = [\theta(Y_0,\xi), \theta(Y_1,\xi), \ldots, \theta(Y_{N_x},\xi)]^T,$$

$$G = [\phi(Y_0,\xi), \phi(Y_1,\xi), \ldots, \phi(Y_{N_x},\xi)]^T$$

are the vector functions at the collocation points. Higher order derivatives are obtained as powers of D, that is

$$f^{(p)} \rightarrow \mathbf{D}^p F, \ u^{(p)} \rightarrow \mathbf{D}^p U, \ h^{(p)} \rightarrow \mathbf{D}^p H, \ \theta^{(p)} \rightarrow \mathbf{D}^p Q, \ \phi^{(p)} \rightarrow \mathbf{D}^p G, \quad (30)$$

where P is the order of the derivative. The grid points on (η, ξ) are defined as

$$Y_j = \cos\frac{\pi j}{N_x}, \quad \xi^n = n\Delta\xi, \quad j=0,1,\ldots,N_x; \ n=0,1,\ldots,N_t,$$

(31)

where $N_x + 1$, $N_t + 1$ are the total number of grid points in the η and ξ-directions respectively, and $\Delta\xi$ is the spacing in the ξ-direction. The finite difference scheme is applied with centering about a mid-point halfway between ξ^{n+1} and ξ^n. This mid-point is defined as $\xi^{n+\frac{1}{2}} = \left(\xi^{n+1} + \xi^n\right)/2$. The derivatives with respect with η are defined in terms of the Chebyshev differentiation matrices.

Applying the centering about $\xi^{n+\frac{1}{2}}$ to any function, say $u(\eta,\xi)$ and its associated derivatives we obtain,

$$u(\eta_j,\zeta^{n+\frac{1}{2}}) = u_j^{n+\frac{1}{2}} = \frac{u_j^{n+1} + u_j^n}{2}, \quad \left(\frac{\partial u}{\partial \xi}\right)^{n+\frac{1}{2}} = \frac{u_j^{n+1} - u_j^n}{\Delta \xi}. \tag{32}$$

Thus, applying the spectral collocation method and finite difference approximation on the SRM scheme (19 – 26) gives

$$\mathbf{A}_1 U_{r+1}^{n+1} = \mathbf{B}_1 U_{r+1}^n + \mathbf{R}_1, \tag{33}$$

$$\mathbf{A}_2 H_{r+1}^{n+1} = \mathbf{B}_2 H_{r+1}^n + \mathbf{R}_2, \tag{34}$$

$$\mathbf{A}_3 Q_{r+1}^{n+1} = \mathbf{B}_3 Q_{r+1}^n + \mathbf{R}_3, \tag{35}$$

$$\mathbf{A}_4 G_{r+1}^{n+1} = \mathbf{B}_4 G_{r+1}^n + \mathbf{R}_4, \tag{36}$$

$$\mathbf{D} F_{r+1}^{n+1} = U_{r+1}^{n+1}, \tag{37}$$

subject to the following boundary and initial conditions

$$u_{r+1}(\eta_0,\zeta^n) = 0, \quad u_{r+1}(\eta_{N_x},\zeta^n) = 1,$$

$$h_{r+1}(\eta_0,\zeta^n) = 0, \quad h_{r+1}(\eta_{N_x},\zeta^n) = 0, \tag{38}$$

$$Q_{r+1}(\eta_0,\zeta^n) = 0, \quad Q_{r+1}(\eta_{N_x},\zeta^n) = 1, \quad f_{r+1}(\eta_{N_x},\zeta^n) = 0, \tag{39}$$

$$G_{r+1}(\eta_0,\zeta^n) = 0, \quad G_{r+1}(\eta_{N_x},\zeta^n) = 1, \tag{40}$$

for $n=0,1,2,\ldots,N_t+1$. The matrices $\mathbf{A}_w, \mathbf{B}_w, \mathbf{R}_w$ are defined for

$$\mathbf{A}_1 = \frac{1}{2}\left(\mathbf{D}^2 + \mathbf{a}_{1,r}^{n+\frac{1}{2}}\mathbf{D}\right) - \frac{\xi^{n+\frac{1}{2}}(1-\xi^{n+\frac{1}{2}})}{\Delta \xi}\mathbf{I},$$

$$\mathbf{B}_1 = -\frac{1}{2}\left(\mathbf{D}^2 + \mathbf{a}_{1,r}^{n+\frac{1}{2}}\mathbf{D}\right) - \frac{\xi^{n+\frac{1}{2}}(1-\xi^{n+\frac{1}{2}})}{\Delta \xi}\mathbf{I}, \quad \mathbf{R}_1 = -\mathbf{a}_{2,r}^{n+\frac{1}{2}},$$

$$A_2 = \frac{1}{2}\left(D^2 + b_{1,r}^{n+\frac{1}{2}}D + b_{2,r}^{n+\frac{1}{2}}\right) - \frac{\xi^{n+\frac{1}{2}}(1-\xi^{n+\frac{1}{2}})}{\Delta\xi}I,$$

$$B_2 = -\frac{1}{2}\left(D^2 + b_{1,r}^{n+\frac{1}{2}}D + b_{2,r}^{n+\frac{1}{2}}\right) - \frac{\xi^{n+\frac{1}{2}}(1-\xi^{n+\frac{1}{2}})}{\Delta\xi}I, \quad R_2 = -b_{3,r}^{n+\frac{1}{2}},$$

$$A_3 = \frac{1}{2}\left(\frac{1}{Pr}D^2 + c_{1,r}^{n+\frac{1}{2}}D\right) - \frac{\xi^{n+\frac{1}{2}}(1-\xi^{n+\frac{1}{2}})}{\Delta\xi}I,$$

$$B_3 = -\frac{1}{2}\left(\frac{1}{Pr}D^2 + c_{1,r}^{n+\frac{1}{2}}D\right) - \frac{\xi^{n+\frac{1}{2}}(1-\xi^{n+\frac{1}{2}})}{\Delta\xi}I, \quad R_3 = O,$$

$$A_4 = \frac{1}{2}\left(\frac{1}{Sc}D^2 + c_{1,r}^{n+\frac{1}{2}}D + d_{1,r}^{n+\frac{1}{2}}\right) - \frac{\xi^{n+\frac{1}{2}}(1-\xi^{n+\frac{1}{2}})}{\Delta\xi}I,$$

$$B_4 = -\frac{1}{2}\left(\frac{1}{Sc}D^2 + c_{1,r}^{n+\frac{1}{2}}D + d_{1,r}^{n+\frac{1}{2}}\right) - \frac{\xi^{n+\frac{1}{2}}(1-\xi^{n+\frac{1}{2}})}{\Delta\xi}I, \quad R_4 = O,$$

where I is an $(N_x+1) \times (N_x+1)$ identity matrix and O is an $(N_x+1)\times 1$ matrix of zeros. The boundary conditions are imposed on the first and last rows of equation each matrix A_w, B_w, R_w. Thus, starting from the initial conditions $U_{r+1}^0, F_{r+1}^0, H_{r+1}^0, Q_{r+1}^0, G_{r+1}^0$ given byequations (27) and (28), the matrix equations (33 – 37) can be solved iteratively, in turn, to give approximate solutions for $u_r +1(\eta,\xi)$, $f_r +1(\eta,\xi)$, etc, for r=0,1,2,..., until a solution that converges to within a given accuracy level is obtained.

RESULTS AND DISCUSSION

In order to determine the evolution of the boundary layer flow properties, numerical solutions of the set of governing systems of partial diff   erential equations (10) – (13) along with the boundary conditions (14), were computed using the proposed spectral relaxation method (SRM). Starting from the initial analytical solutions at $\xi=0$ (corresponding to $\tau=0$), the SRM scheme was used to generate results up to solutions near the steady state values at $\xi = 1$ (corresponding to $\tau \rightarrow \infty$). The effect of the governing parameters namely, the rotation rate parameter λ, the Schmidt number Sc, the non-dimensional activation energy E, the Prandtl number Pr, the chemical reaction rate constant σ, the temperature relative parameter d and n on the flow characteristics as well as the local skin friction, heat and mass transfer coefficients the results are presented graphically in this section. Fig. 1 and Fig. 2 show the variation of the velocity profiles $h(\eta)$ and $f'(\eta)$, respectively, for different values of λ. We observe that an increase in the values of λ leads to monotonic exponential decay in the velocity profiles for small values and it results in oscillatory decay for a large values of λ. The same results have been reported by Nasar et al. [14] in a related study. Fig. 3 and Fig. 4 show the variation of the skin friction coefficients in the x and y directions respectively for various values of the rotation rate parameter λ. It is observed that I decreases both the skin friction coefficients thus reduces the momentum boundary layers. The effects of the rotation rate parameter λ on the temperature profile is shown in Fig. 5. This figure shows that the thermal boundary layer thickness decreases with λ, thus an increase in λ causing a drop in the temperature. Fig. 6 illustrates the variation of the Nusselt number $Nux=Re1=2$ x with j for some values of λ. However increases λ decreases the heat transfer coefficient and the influence of λ can be obtained beyond $\xi \geq 0.4$

in the heat. The variations of the temperature $\theta(\eta, \xi)$ profile with g for several values of the Prandtl number Pr are shown in Fig. 7. It is observed that the thermal boundary layer thickness decrease with an increase in Pr. Larger values of Prandtl number corresponds to the weaker thermal diffusivity and thinner boundary layer, hence Pr reduces the temperature. Fig. 8 shows concentration distribution for several values Prandtl number. The effect of the Prandtl number is to reduce the mass transfer boundary-layer thickness and so reducing

the $\phi(\eta, \xi)$ The influence of the chemical reaction rate constant σ on the concentration profile within the boundary layer is given in Fig. 9. An increase in the σ effect reduces the concentration within the thermal boundary layer region. This is because increasing the chemical reaction rate causes a thickening of the mass transfer boundary layer. The effects of the non-dimensional activation energy E on the concentration profile have been plotted in Fig. 10, it has been notice that increasing the non-dimensional activation energy E effect increases the concentration boundary layer thinness which enhances the concentration.

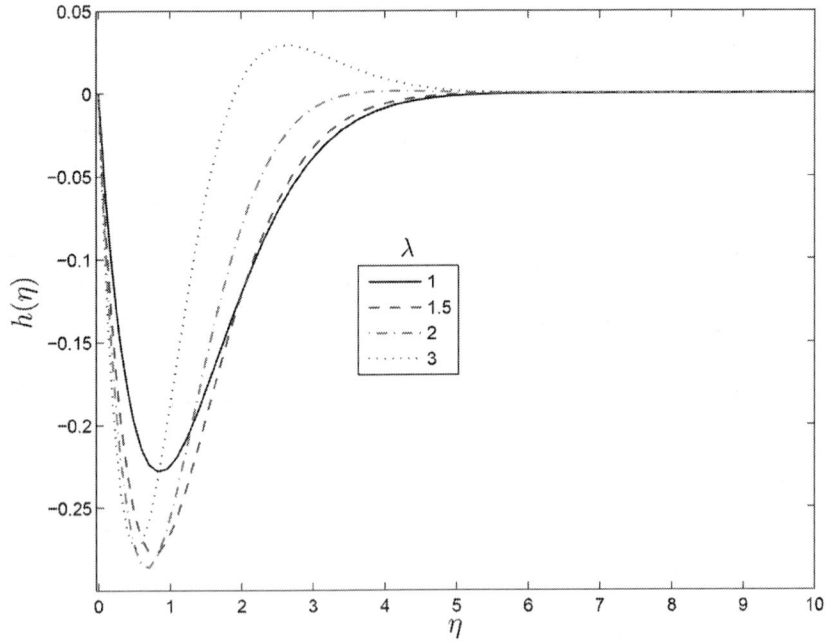

Figure 1: Effect of the rotating parameter λ on $h(\eta)$ for $\xi = 0.65$, Sc=1, $\sigma = 5$, Pr=0.71, E=1, $\delta = 1$ and n=0.5.

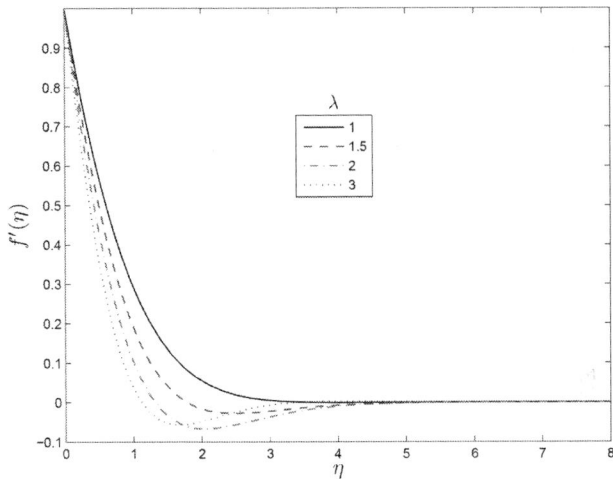

Figure 2: Effect of the rotating parameter λ on f'(n) for ξ =0.65, Sc=1, σ =5, Pr=0.71, E=1, δ =1 and n=0.5.

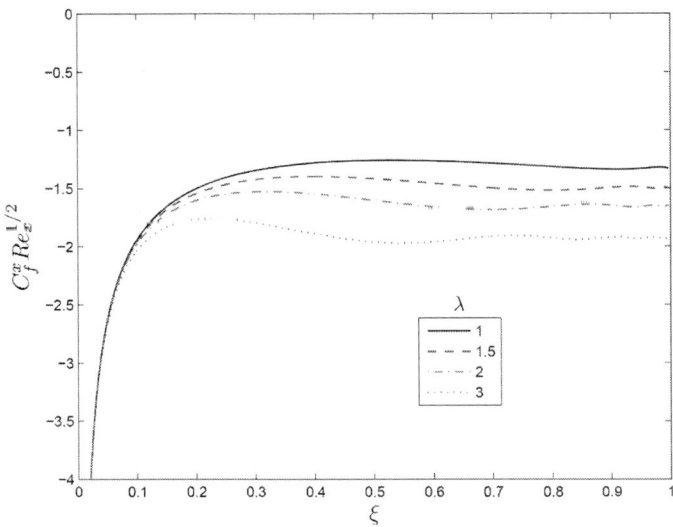

Figure 3: Effect of the rotating parameter λ on $C_f^x Re^{1/2}$ for, ξ =0.65, Sc=1, σ =5, Pr=0.71, E=1, δ =1 and n=0.5.

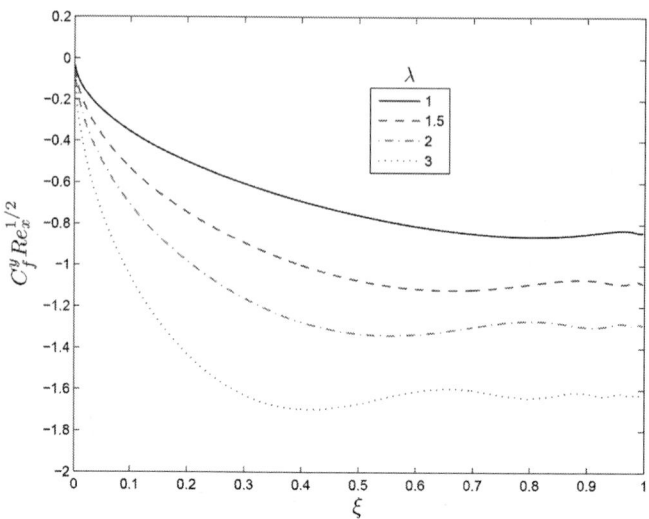

Figure 4: Effect of the rotating parameter λ on $C_f^y Re^{1/2}$ for, $\xi = 0.65$, Sc=1, $\sigma = 5$, Pr=0.71, E=1, $\delta = 1$ and n=0.5.

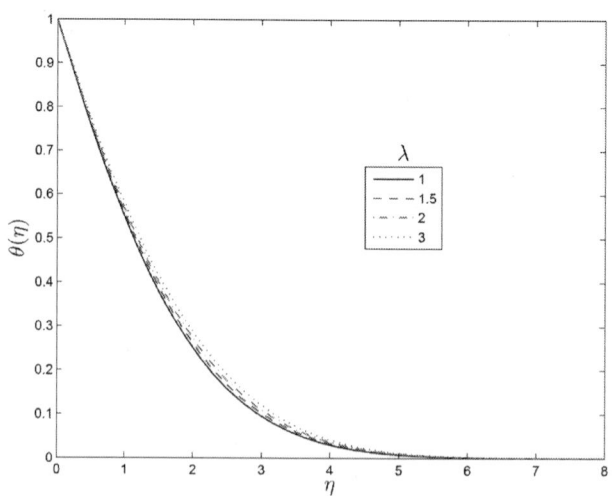

Figure 5: Effect of the rotating parameter λ on $\theta(\eta)$ for, $\xi = 0.65$, Sc=1, $\sigma = 5$, Pr=0.71, E=1, $\delta = 1$ and n=0.5.

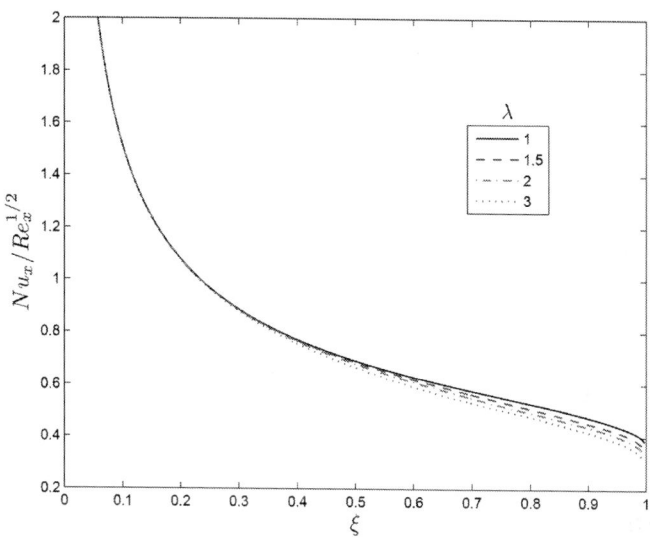

Figure 6: Effect of the rotating parameter λ on $Nu_x / Re_x^{1/2}$ for, ξ =0.65, Sc=1, σ =5, Pr=0.71, E=1, δ =1 and n=0.5.

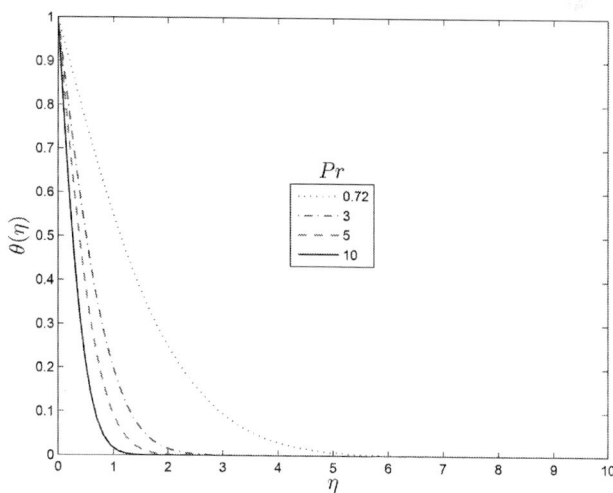

Figure 7: Effect of the rotating parameter Pr on $\theta(\eta)$ for, ξ =0.65, Sc=1, σ =5, Pr=0.71, E=1, δ =1 and n=0.5.

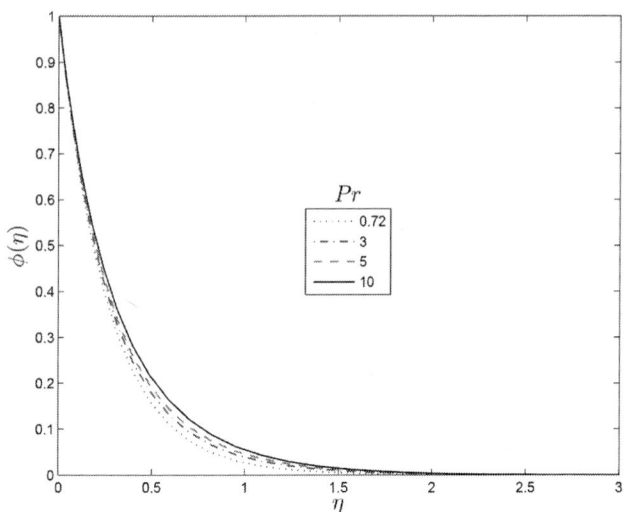

Figure 8: Effect of the rotating parameter Pr on $\theta(\eta)$ for, ξ =0.65, Sc=1, σ =5, Pr=0.71, E=1, δ =1 and n=0.5.

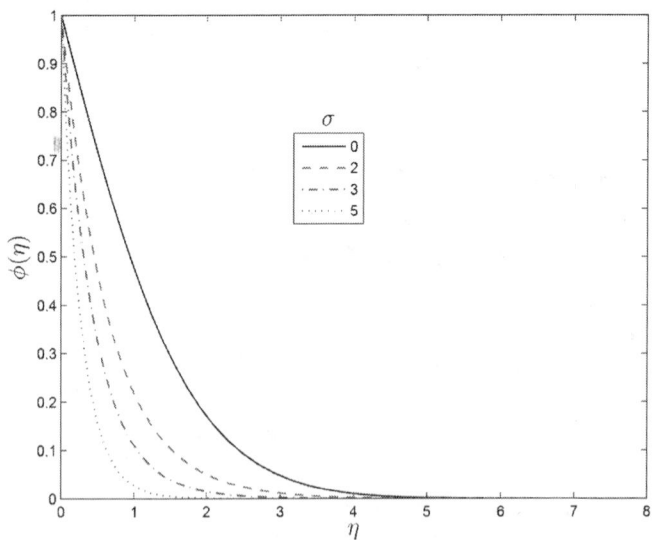

Figure 9: Effect of σ **on** $\theta(\eta)$ **for** ξ =0.65, E=1, Sc=1, λ =5, Pr=0.71, δ =1 and n=0.5.

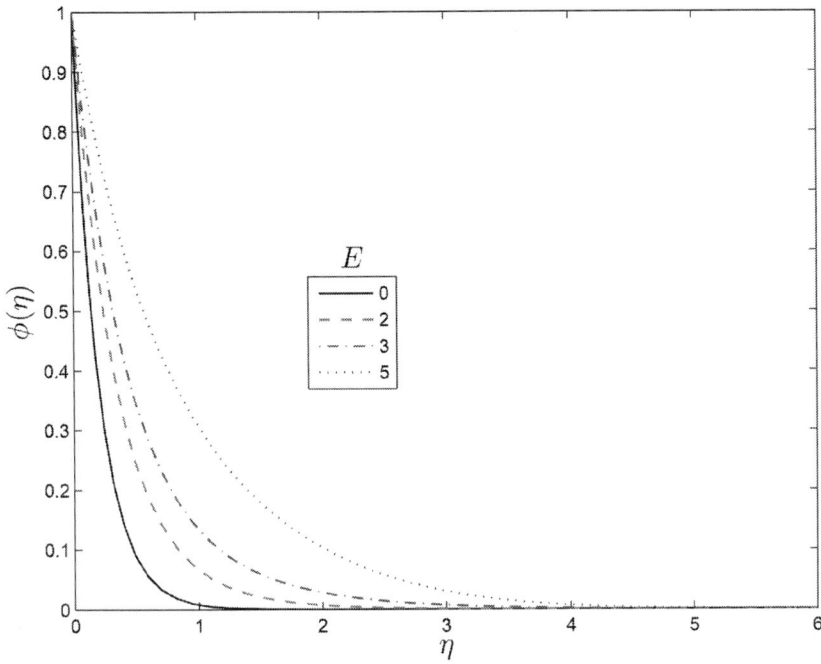

Figure 10: Effect of E on $\phi(\eta)$ for ξ =0.65, σ =5, Sc=1, λ =5, Pr=0.71, δ =1 and n=0.5.

Fig. 11 shows the effect of increasing the dimensionless exponent fitted rate constant n on the concentration profile. It is observed that increasing n reduces the concentration within the thermal boundary layer leading to an increase in the concentration gradient at the sheet. FromFig. 12 dimensionless exponent fitted rate constant n leads to a considerable thinning of the concentration boundary layer, and hence a reduction in mass transfer rate at the sheet wall.Fig. 13 and Fig. 14 depict the variation of the solute concentration and the mass transfer rate $Shx / Re_x^{1/2}$ respectively for different values of the temperature relative parameter δ . It is evident that as δ increases, the concentration boundary layer thickness decreases followed by a reduction in both the solute concentration and the mass transfer rate.

Figure 11: Effect of the rotating parameter n on $\phi(\eta)$ for ξ =0.65, Sc=1, λ =1, Pr=0.71, σ =5, δ =1 and E=1

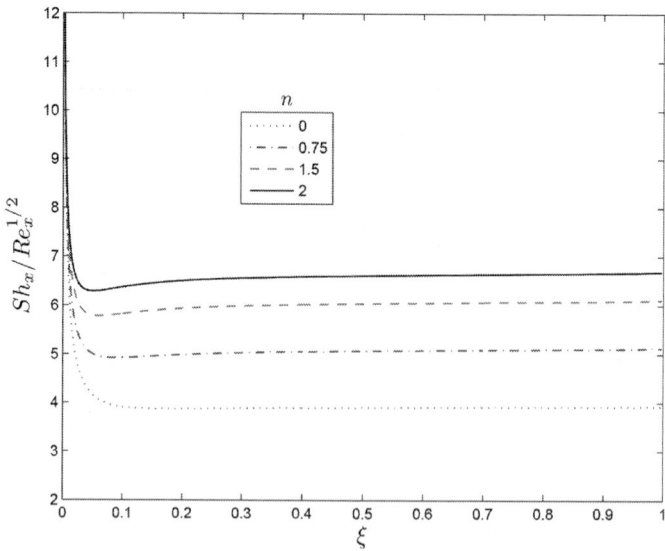

Figure 12: Effect of the rotating parameter n on $Shx / Re_x^{1/2}$ for, ξ =0.65, Sc=1, λ =1, Pr=0.71, σ =5, δ =1 and E=1.

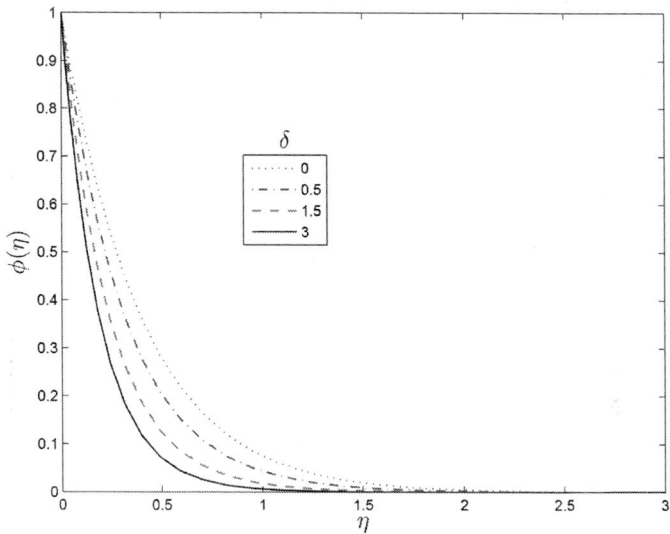

Figure 13: Effect of the rotating parameter n on $\phi(\eta)$ for ξ =0.65, Sc=1, λ =1, Pr=0.71, σ =5, E=1 and n=0.5

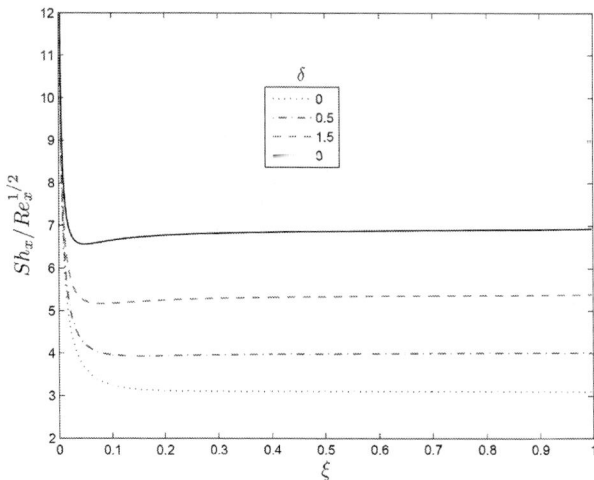

Figure 14: Effect of the rotating parameter δ on $Shx/Re_x^{1/2}$ for, ξ =0.65, Sc=1, λ =1, Pr=0.71, σ =5, E=1 and n=0.5.

CONCLUSIONS

In this investigation, we considered the spectral relaxation method approach to solving an coupled non-linear partial differential equation system that governs the unsteady flow with binary chemical reaction and activation energy due to a stretching surface in a rotating fluid. The effects of the governing parameters namely the rotation rate parameter, the Schmidt number, the non-dimensional activation energy, the Prandtl number, the chemical reaction rate constant, the temperature relative parameter and on the flow characteristics as well as the local skin friction, heat and mass transfer coefficients have been studied. Small values the rotation rate parameter λ shows a monotonic exponential decay in the velocity profiles and there is oscillatory decay for a large values. Increasing in the non-dimensional activation energy E enhances the concentration profile within the boundary layer. The spectral relaxation method used was found to be a very effective method for solving the type of problem considered in this work.

AUTHOR CONTRIBUTIONS

Conceived and designed the experiments: FA. Analyzed the data: SM FA. Contributed reagents/materials/analysis tools: FA MK SM. Wrote the paper: FA SM. Provided financial support: MK.

REFERENCES

1.   Sakiadis BC (1961) Boundary layer behaviour on continuous solid surface: I. Boundary layer equations for two-dimensional and axismmetric flow. AIChE Journal 7: 26–28. doi: 10.1002/aic.690070108

2.   Sakiadis BC (1961) Boundary layer behaviour on continuous solid surface: II. Boundary layer equations for two-dimensional and axismmetric flow. AIChE Journal 7: 221–225. doi: 10.1002/aic.690070211

3.   Crane LJ (1970) Flow past a stretching plate, Zeitschrift für angewandte Mathematik und physik. 12: 645–647. doi: 10.1007/bf01587695

4.   Brady JF, Acrivos A (1981) Steady flow in a channel or tube with an accelerating surface velocity. J. Fluid Mech. 112 (1981): 127–150.

5.   Wang CY (1984) The three-dimensional flow due to a stretching flat surface. Phys. Fluids 27: 1915–1917. doi: 10.1063/1.864868

6.   Afzal N (1993) Heat transfer froma stretching surface, International Journal of Heat and Mass Transfer. 36: 1128–1131. doi: 10.1016/s0017-9310(05)80296-0

7.   Prasad KV, Abel S, Datti PS (2003) Diffusion of chemically reactive species of a non-Newtonian fluid immersed in a porous medium over a stretching sheet. International Journal of Non- Linear Mechanics 38: 651–657. doi: 10.1016/s0020-7462(01)00122-6

8.   Abel MS, Siddheshwar PG, Nandeppanavar MM (2007) Heat transfer in a viscoelastic boundary layer flow over a stretching sheet with viscous dissipation and non-uniform heat source. International Journal of Heat andMass Transfer: 50: 960–966.

9.   Bataller RC (2007) Viscoelastic fluid flow and heat transfer over a stretching sheet under the effects of a non-uniform heat source, viscous dissipation and thermal radiation, International Journal of Heat and Mass Transfer. 50: 3152–3162. doi: 10.1016/j.ijheatmasstransfer.2007.01.003

10.   Abel MS, Mahesha N (2008) Heat transfer in MHD viscoelastic fluid flow over a stretching sheet with variable thermal conductivity, non-uniform heat source and radiation, Applied Mathematical Modelling, 32 (2008): 1965–1983.

11.   Soundalgekar VM, Martin BW, Gupta SK, Pop I (1976) On Unsteady Boundary Layer in a Rotating Fluid with Time dependant suction, Publication De L'Institute Mathematique. 20: 215–226.

12. Bergstrom RW (1976) Viscous Boundary Layers in Rotating Fluids Driven by Periodic Flows. J. Atmos. Sci 33: 1234–1247. doi: 10.1175/1520-0469(1976)033<1234:vblirf>2.0.co;2

13. Abbas Z, Javed T, Sajid M, Ali N (2010) Unsteady MHD flow and heat transfer on a stretching sheet in a rotating fluid, Journal of the Taiwan Institute of Chemical Engineers. 41: 644–650. doi: 10.1016/j.jtice.2010.02.002

14. Nazar R, Amin N, Pop I (2004) Unsteady boundary layer flow due to a stretching surface in a rotating fluid, Mechanics Research Communications. 31: 121–128. doi: 10.1016/j.mechrescom.2003.09.004

15. Zheng L, Li C, Zhang X, Gao Y (2011) Exact solutions for the unsteady rotating flows of a generalized Maxwell fluid with oscillating pressure gradient between coaxial cylinders, Computers & Mathematics with Applications. 62: 1105–1115. doi: 10.1016/j.camwa.2011.02.044

16. Fang T, Tao H (2012) Unsteady viscous flow over a rotating stretchable disk with deceleration, Communications in Nonlinear Science and Numerical. Simulation17: 5064–5072. doi: 10.1016/j.cnsns.2012.04.017

17. Rashad AM (2014) Effects of radiation and variable viscosity on unsteady MHD flow of a rotating fluid from stretching surface in porous media, Journal of the Egyptian Mathematical Society. 22: 134–142. doi: 10.1016/j.joems.2013.05.008

18. Khan NA, Aziz S, Khan NA (2014) Numerical simulation for the unsteady MHD flow and heat transfer of couple stress fluid over a rotating disk, PLoS ONE. 9(5): e95423 doi:10.1371/journal.pone.0095423.

19. Hajmohammadi MR, Nourazar SS, Campo A (2014) Analytical solution for two-phase flow betwen two rotating cylinders filed with power law liquid and a micro layer of gas, Journal of Mechanical Science and Technology. 28: 1849–1854. doi: 10.1007/s12206-014-0332-y

20. Hajmohammadi MR, Nourazar SS (2014) On the insertion of a thin gas layer in micro cylindrical Couette flows

involving power-law liquids International Journal of Heat and Mass Transfer. 75: 97–108. doi: 10.1016/j.ijheatmasstransfer.2014.03.065

21. Bestman AR (1990) Natural convection boundary layer with suction and mass transfer in a porous medium, International Journal of Energy Research. 14: 389–396. doi: 10.1002/er.4440140403

22. Bestman AR (1991) Radiative heat transfer to flow of a combustible mixture in a vertical pipe, International Journal of Energy Research. 15: 179–184. doi: 10.1002/er.4440150305

23. Makinde OD, Olanrewaju PO, Charles WM (2011) Unsteady convection with chemical reaction and radiative heat transfer past a flat porous plate moving through a binary mixture, Africka Matematika. 22: 65–78. doi: 10.1007/s13370-011-0008-z

24. Makinde OD, Olanrewaju PO (2011) Unsteady mixed convection with Soret and Dufour effects past a porous plate moving through a binary mixture of chemically reacting fluid, Chemical Engineering Communications. 198: 920–938. doi: 10.1080/00986445.2011.545296

25. AbdulMaleque Kh (2013) Effects of binary chemical reaction and activation energy on MHD boundary layer heat and mass transfer flow with viscous dissipation and heat generation/absorption, Hindawi Publishing Corporation, ISRN Thermodynamics Volume 2013: Article ID 284637,doi:10.1155/2013/284637.

26. AbdulMaleque Kh (2013) Unsteady natural convection boundary layer flow with mass transfer and a binary chemical reaction, Hindawi Publishing Corporation, British Journal of Applied Science & Technology volume. 2013: 131–149. doi: 10.9734/bjast/2014/2265

27. AbdulMaleque Kh (2013) Effects of exothermic/endothermic chemical reactions with arrhenius activation energy on MHD free convection and mass transfer flow in presence of thermal radiation, Hindawi Publishing Corporation,

Journal ofThermodynamics Volume 2013: Article ID 692516, doi:10.1155/2013/692516.

28.   Canuto C, Hussaini MY, Quarteroni A, Zang TA (1988) Spectral Methods in Fluid Dynamics, Springer-Verlag, Berlin.

29.   Motsa SS, Makukula ZG (2013) On spectral relaxation method approach for steady von Karman flow of a Reiner-Rivlin fluid with Joule heating, vicious dissipation and suction/injection. Cent. Eur. J. Phys 11: 363–374.. doi: 10.2478/s11534-013-0182-8

30.   Motsa SS (2014) A new spectral relaxation method for similarity variable nonlinear boundary layer flow systems, Chemical Engineering Communications. 201: 241–256. doi: 10.1080/00986445.2013.766882

31.   Motsa SS, Dlamini PG, Khumalo M (2012) Solving Hyperchaotic Systems Using the Spectral Relaxation Method, Abstract and Applied Mathematics Volume 2012:, Article ID 203461, 18 pages doi:10.1155/2012/203461.

32.   Motsa SS, Dlamini PG, Khumalo M (2013) A new multistage spectral relaxation method for solving chaotic initial value systems, Nonlinear Dyn. 72: 265–283. doi: 10.1007/s11071-012-0712-8

Citations

CHAPTER 1

Mohammad Songolzadeh, Mansooreh Soleimani, Maryam Takht Ravanchi, and Reza Songolzadeh, "Carbon Dioxide Separation from Flue Gases: A Technological Review Emphasizing Reduction in Greenhouse Gas Emissions," The Scientific World Journal, vol. 2014, Article ID 828131, 34 pages, 2014. doi:10.1155/2014/828131.

CHAPTER 2

Rutuja Bhoje, Ganesh R. Kale, Nitin Labhsetwar, and Sonali Borkhade, "Chemical Looping Combustion of Methane: A

Technology Development View," Journal of Energy, vol. 2013, Article ID 949408, 15 pages, 2013. doi:10.1155/2013/949408.

CHAPTER 3

Colin Awungacha Lekelefac, Nadine Busse, Michael Herrenbauer, and Peter Czermak, "Photocatalytic Based Degradation Processes of Lignin Derivatives," International Journal of Photoenergy, vol. 2015, Article ID 137634, 18 pages, 2015. doi:10.1155/2015/137634.

CHAPTER 4

Jean-Claude Charpentier, Four main objectives for the future of chemical and process engineering mainly concerned by the science and technologies of new materials production, Chemical Engineering Journal, Volume 107, Issues 1–3, 15 March 2005, Pages 3-17, ISSN 1385-8947, http://dx.doi.org/10.1016/j.cej.2004.12.004.

CHAPTER 5

Kazuhiro Mae, Advanced chemical processing using microspace, Chemical Engineering Science, Volume 62, Issues 18–20, September–October 2007, Pages 4842-4851, ISSN 0009-2509, http://dx.doi.org/10.1016/j.ces.2007.01.012.

CHAPTER 6

Awad FG, Motsa S, Khumalo M (2014) Heat and Mass Transfer in Unsteady Rotating Fluid Flow with Binary Chemical Reaction and Activation Energy. PLoS ONE 9(9): e107622. doi:10.1371/journal.pone.0107622.

Index